AF158391

The publishing house tredition has created the series **TREDITION CLASSICS**. It contains classical literature works from over two thousand years. Most of these titles have been out of print and off the bookstore shelves for decades.

The book series is intended to preserve the cultural legacy and to promote the timeless works of classical literature. As a reader of a **TREDITION CLASSICS** book, the reader supports the mission to save many of the amazing works of world literature from oblivion.

The symbol of **TREDITION CLASSICS** is Johannes Gutenberg (1400 – 1468), the inventor of movable type printing.

With the series, tredition intends to make thousands of international literature classics available in printed format again – worldwide.

All books are available at book retailers worldwide in paperback and in hardcover. For more information please visit: www.tredition.com

tredition was established in 2006 by Sandra Latusseck and Soenke Schulz. Based in Hamburg, Germany, tredition offers publishing solutions to authors and publishing houses, combined with worldwide distribution of printed and digital book content. tredition is uniquely positioned to enable authors and publishing houses to create books on their own terms and without conventional manufacturing risks.

For more information please visit: www.tredition.com

Researches on Cellulose 1895-1900

C. F. Cross

Imprint

This book is part of the TREDITION CLASSICS series.

Author: C. F. Cross
Cover design: toepferschumann, Berlin (Germany)

Publisher: tredition GmbH, Hamburg (Germany)
ISBN: 978-3-8495-0993-4

www.tredition.com
www.tredition.de

Copyright:
The content of this book is sourced from the public domain.

The intention of the TREDITION CLASSICS series is to make world literature in the public domain available in printed format. Literary enthusiasts and organizations worldwide have scanned and digitally edited the original texts. tredition has subsequently formatted and redesigned the content into a modern reading layout. Therefore, we cannot guarantee the exact reproduction of the original format of a particular historic edition. Please also note that no modifications have been made to the spelling, therefore it may differ from the orthography used today.

PREFACE TO SECOND EDITION

This edition is a *reprint* of the first in response to a continuous demand for the book. The matter, consisting as it does largely of records, does not call for any revision, and, as a contribution to the development of theory, any particular interest which it has is associated with the date at which it was written.

The volume which has since appeared is the sequel, and aims at an exposition of the subject "to date".

PREFACE

This volume, which is intended as a supplement to the work which we published in 1895, gives a brief account of researches which have been subsequently published, as well as of certain of our own investigations, the results of which are now for the first time recorded.

We have not attempted to give the subject-matter the form of a connected record. The contributions to the study of 'Cellulose' which are noticed are spread over a large area, are mostly 'sectional' in their aim, and the only cohesion which we can give them is that of classifying them according to the plan of our original work. Their subject-matter is reproduced in the form of a *précis*, as much condensed as possible; of the more important papers the original title is given. In all cases we have endeavoured to reproduce the Author's main conclusions, and in most cases without comment or criticism.

Specialists will note that the basis of investigation is still in a great measure empirical; and of this the most obvious criterion is the confusion attaching to the use of the very word 'Cellulose.' This is due to various causes, one of which is the curious specialisation of the term in Germany as the equivalent of 'wood cellulose.' The restriction of this general or group term has had an influence even in scientific circles. Another influence preventing the recognition of the obvious and, as we think, inevitable basis of classification of the 'celluloses' is the empiricism of the methods of agricultural chemistry, which as regards cellulose are so far chiefly concerned with its negative characteristics and the analytical determination of the indigestible residue of fodder plants. Physiologists, again, have their own views and methods in dealing with cellulose, and have hitherto had but little regard to the work of the chemist in differentiating and classifying the celluloses on a systematic basis. There are many sides to the subject, and it is only by a sustained effort towards centralisation that the general recognition of a systematic basis can be secured.

We may, we hope usefully, direct attention to the conspicuous neglect of the subject in this country. To the matter of the present volume, excluding our own investigations, there are but two contributions from English laboratories. We invite the younger generation of students of chemistry to measure the probability of finding a working career in connection with the cellulose industries. They will not find this invitation in the treatment accorded to the subject in text-books and lectures. It is probable, indeed, that the impression produced by their studies is that the industries in coal-tar products largely exceed in importance those of which the carbohydrates are the basis; whereas the former are quite insignificant by comparison. A little reflection will prove that cellulose, starch, and sugar are of vast industrial moment in the order in which they are mentioned. If it is an open question to what extent science follows industry, or *vice versa*, it is not open to doubt that scientific men, and especially chemists, are called in these days to lead and follow where industrial evolution is most active. There is ample evidence of activity and great expansion in the cellulose industries, especially in those which involve the chemistry of the raw material; and the present volume should serve to show that there is rapid advance in the science of the subject. Hence our appeal to the workers not to neglect those opportunities which belong to the days of small beginnings.

We have especially to acknowledge the services of Mr. J. F. Briggs in investigations which are recorded on pp. 34-40 and pp. 125-133 of the text.

CONTENTS

THE MATTER OF THIS VOLUME MAY BE DIVIDED INTO THE FOLLOWING SECTIONS

INTRODUCTION – DEALING WITH THE SUBJECT IN GENERAL OUTLINE

SECTION

I. GENERAL CHEMISTRY OF THE TYPICAL COTTON CELLULOSE

II. SYNTHETICAL DERIVATIVES – SULPHOCARBONATES AND ESTERS

III. DECOMPOSITIONS OF CELLULOSE SUCH AS THROW LIGHT ON THE PROBLEM OF ITS CONSTITUTION

IV. CELLULOSE GROUP, INCLUDING HEMICELLULOSES AND TISSUE CONSTITUENTS OF FUNGI

V. FURFUROIDS, *i.e.* PENTOSANES AND FURFURAL-YIELDING CONSTITUENTS GENERALLY

VI. THE LIGNOCELLULOSES

VII. PECTIC GROUP

VIII. INDUSTRIAL AND TECHNICAL. GENERAL REVIEW

INDEX OF AUTHORS

INDEX OF SUBJECTS

CELLULOSE

INTRODUCTION

In the period 1895-1900, which has elapsed since the original publication of our work on 'Cellulose,' there have appeared a large number of publications dealing with special points in the chemistry of cellulose. So large has been the contribution of matter that it has been considered opportune to pass it under review; and the present volume, taking the form of a supplement to the original work, is designed to incorporate this new matter and bring the subject as a whole to the level to which it is thereby to be raised. Some of our critics in reviewing the original work have pronounced it 'inchoate.' For this there are some explanations inherent in the matter itself. It must be remembered that every special province of the science has its systematic beginning, and in that stage of evolution makes a temporary 'law unto itself.' In the absence of a dominating theory or generalisation which, when adopted, gives it an organic connection with the general advance of the science, there is no other course than to classify the subject-matter. Thus 'the carbohydrates' may be said to have been in the inchoate condition, qualified by a certain classification, prior to the pioneering investigations of Fischer. In attacking the already accumulated and so far classified material from the point of view of a dominating theory, he found not only that the material fell into systematic order [Pg 2] and grew rapidly under the stimulus of fruitful investigation, but in turn contributed to the firmer establishment of the theoretical views to which the subject owed its systematic new birth. On the other hand, every chemist knows that it is only the simpler of the carbohydrates which are so individualised as to be connoted by a particular formula in the stereoisomeric system. Leaving the monoses, there is even a doubt as to the constitution of cane sugar; and the elements of uncertainty thicken as we approach the question of the chemical structure of starch. This unique product of plant life has a literature of its own, and how little of this is fully known to what we may term the 'average chemist' is seen by the methods he will employ for its quantitative estimation. In one particular review of our work where we are taken to task for producing 'an aggravating book, inchoate in the highest degree ... disfigured by an obscurity of diction which must materially diminish its usefulness' ['Nature,' 1897, p. 241], the author, who is a well-known and competent critic, makes use of the short expression in regard to the more complex carbohydrates, 'Above cane sugar, higher in the series, all is chaos,' and in reference to starch, 'the subject is still enshrouded in mystery.' This 'material' complexity is at its maximum with the most complex members of the series, which are the celluloses, and we think accounts in part for the impatience of our critic. 'Obscurity of diction' is a personal quantity, and we must leave that criticism to the fates. We find also that many workers whose publications we notice in this present volume quite ignore the *plan* of the work, though they make use of its matter. We think it necessary to restate this plan, which, we are satisfied, is systematic, and, in fact, inevitable. Cellulose is in the first instance a *structure*, and the anatomical relationships supply a certain basis of classification. Next, it is known to us and is defined [Pg 3] by the negative characteristics of resistance to hydrolytic actions and oxidations. These are dealt with in the order of their intensity. Next we have the more positive definition by ultimate products of hydrolysis, so far as they are known, which discloses more particularly the presence of a greater or less proportion of furfural-yielding groups. Putting all these together as criteria of function and composition we find they supply common or general dividing lines, within which groups of these products are contained. The classification is natural, and in that sense inevitable; and it not only groups the physiological and chemical facts, but the industrial also. We do not propose to argue the question whether the latter adds any cogency to a scientific scheme. We are satisfied that it does, and we do not find any necessity to exclude a particular set of phenomena from consideration, because they involve 'commercial' factors. We have dealt with this classification in the original work (p. 78), and we discuss its essential basis in the present volume (p. 28) in connection with the definition of a 'normal' cellulose. But the 'normal' cellulose is not the only cellulose, any more than a primary alcohol or an aliphatic alcohol are the only alcohols. This point is confused or ignored in several of the recent contributions of investigators. It will suffice to cite one of these in illustration. On p. 16 we give an account of an investigation of the several methods of estimating cellulose, which is full of valuable and interesting matter. The purpose of the author's elaborate comparative study is to decide which has the strongest claims to be regarded as the 'standard' method. They appear to have a preference for the method of Lange—viz. that of heating at high temperatures (180°) with alkaline hydrates, but the investigation shows that (as we had definitely stated in our original work, p. 214) this is subject to large and variable errors. The adverse judgment of the authors, we may [Pg 4] point out, is entirely determined on the question of aggregate weight or yield, and without reference to the ultimate composition or constitution of the final product. None of the available criteria are applied to the product to determine whether it is a cellulose (anhydride) or a hydrate or a hydrolysed product. After these alkali-fusion processes the method of chlorination is experimentally reviewed and dismissed for the reason that the product retains furfural-yielding groups, which is, from our point of view, a particular recommendation, i.e. is evidence of the selective action of the chlorine and subsequent hydrolysis upon the lignone group. As a matter of fact it is the only method yet available for isolating the cellulose from a lignocellulose by a treatment which is quantitatively to be accounted for in every detail of the reactions. It does not yield a 'normal' cellulose, and this is the expression which, in our opinion, the authors should have used. It should have been pointed out, moreover, that, as the cellulose is separated from actual condensed combination with the lignone groups, it may be expected to be obtained in a hydrated form, and also not as a homogeneous substance like the normal cotton cellulose. The product is a cellulose of the second group of the classification. Another point in this investigation which we must criticise is the ultimate selection of the Schulze method of prolonged maceration with nitric acid and a chlorate, followed by suitable hydrolysis of the non-cellulose derivatives to soluble products. Apart from its exceptional inconvenience, rendering it quite impracticable in laboratories which are concerned with the valuation of cellulosic raw materials for industrial purposes, the attack of the reagent is complex and ill-defined. This criticism we would make general by pointing out that such processes quite ignore the specific characteristics of the non-cellulose components of the compound celluloses. The second division of the plan of our [Pg 5] work was to define these constituents by bringing together all that had been established concerning them. These groups are widely divergent in chemical character, as are the compound celluloses in function in the plant. Consequently there is for each a special method of attack, and it is a reversion to pure empiricism to expect any one treatment to act equally on the pectocelluloses, lignocelluloses, and cutocelluloses. Processes of isolating cellulose are really more strictly defined as methods of selective and regulated attack of the groups with which they occur, combined or mixed. A chemist familiar with such types as rhea or ramie (pectocellulose), jute (lignocellulose), and raffia (cutocellulose) knows exactly the specific treatment to apply to each for isolating the cellulose, and must view with some surprise the appearance at this date of such 'universal prescriptions' as the process in question.

The third division of our plan of arrangement comprised the synthetical derivatives of the celluloses, the sulphocarbonates first, as peculiarly characteristic, and then the esters, chiefly the acetates, benzoates, and nitrates. To these, investigators appear to have devoted but little attention, and the contribution of new matter in the present volume is mainly the result of our own researches. It will appear from this work that an exhaustive study of the cellulose esters promises to assist very definitely in the study of constitutional problems.

This brings us to the fourth and, to the theoretical chemist, the most important aspect of the subject, the problem of the actual molecular structure of the celluloses and compound celluloses. It is herein we are of opinion that the subject makes a 'law unto itself.' If the constitution of starch is shrouded in mystery and can only be vaguely expressed by generalising a complex mass of statistics of its successive hydrolyses, we can only still more vaguely guess at [Pg 6] the distance which separates us from a mental picture of the cellulose unit. We endeavour to show by our later investigations that this problem merges into that of the actual structure of cellulose in the mass. It is definitely ascertained that a change in the molecule, or reacting unit, of a cellulose, proportionately affects the structural properties of the derived compounds, both sulphocarbonates and esters. This is at least an indication that the properties of the visible aggregates are directly related to the actual configuration of the chemical units. But it appears that we are barred from the present discussion of such a problem in absence of any theory of the solid state generally, but more particularly of those forms of matter which are grouped together as 'colloids.'

Cellulose is distinguished by its inherent constructive functions, and these functions take effect in the plastic or colloidal condition of the substance. These properties are equally conspicuous in the synthetical derivatives of the compound. Without reference, therefore, to further speculations, and not deterred by any apparent hopelessness of solving so large a problem, it is clear that we have to exhaust this field by exact measurements of all the constants which can be reduced to numerical expression. It is most likely that the issue may conflict with some of our current views of the molecular state which are largely drawn from a study of the relatively dissociated forms of matter. But such conflicts are only those of enlargement, and we anticipate that all chemists look for an enlargement of the molecular horizon precisely in those regions where the forces of cell-life manifest themselves.

The *cellulose group* has been further differentiated by later investigations. The fibrous celluloses of which the typical members receive important industrial applications, graduate by insensible stages into the hemicelluloses which may be regarded as a well-established sub-group. In considering [Pg 7] their morphological and functional relationships it is evident that the graduation accords with their structure and the less permanent functions which they fulfil. They are aggregates of monoses of the various types, chiefly mannose, galactose, dextrose, &c., so far as they have been investigated.

Closely connected with this group are the constituents of the tissues of fungi. The recent researches of Winterstein and Gilson, which are noted in this present volume, have established definitely that they contain a nitrogenous group in intimate combination with a carbohydrate complex. This group is closely related to chitin, yielding glucosamin and acetic acid as products of ultimate hydrolysis. Special interest attaches to these residues, as they are in a sense intermediate products between the great groups of the carbohydrates and proteids (E. Fischer, Ber. 19, 1920), and their further investigation by physiological methods may be expected to disclose a genetic connection.

The *lignocelluloses* have been further investigated. Certain new types have been added, notably a soluble or 'pectic' form isolated from the juice of the white currant (p. 152), and the pith-like wood of the Æschynomene (p. 135).

Further researches on the typical fibrous lignocellulose have given us a basis for correcting some of the conclusions recorded in our original work, and a study of the esters has thrown some light on the constitution of the complex (p. 130).

Of importance also is the identification of the hydroxyfurfurals as constituents of the lignocelluloses generally, and the proof that the characteristic colour-reactions with phenols (phloroglucinol) may be ascribed to the presence of these compounds (p. 116).

The *pectocelluloses* have not been the subject of systematic chemical investigation, but the researches of Gilson ('La Cristallisation de la Cellulose et la Composition Chimique de [Pg 8] la Membrane Cellulaire Végétale,' 'La Revue,' 'La Cellule,' i. ix.) are an important contribution to the natural history of cellulose, especially in relation to the 'pectic' constituents of the parenchymatous celluloses. Indirectly also the researches of Tollens on the 'pectins' have contributed to the subject in correcting some of the views which have had a text-book currency for a long period. These are dealt with on p. 151. The results establish that the pectins are rather the soluble hydrated form of cellulosic aggregates in which acid groups may be represented; but such groups are not to be regarded as essentially characteristic of this class of compounds.

Furfural-yielding Substances (Furfuroids).—This group of plant products has been, by later investigations, more definitely and exclusively connected with the celluloses—i.e. with the more permanent of plant tissues. From the characteristic property of yielding furfural, which they have in common with the pentoses, they have been assumed to be the anhydrides of these C_5 sugars or pentosanes; but the direct evidence for this assumption has been shown to be wanting. In regard to their origin the indirect evidences which have accumulated all point to their formation in the plant from hexoses. Of special interest, in its bearings on this point, is the direct transformation of levulose into furfural derivatives, which takes place under the action of condensing agents. The most characteristic is that produced by the action of anhydrous hydrobromic acid in presence of ether [Fenton], yielding a brommethyl furfural

$C_6H_{12}O_6 - 4H_2O + HBr = C_5H_3.O_2.CH_2Br$

with a Br atom in the methyl group. These researches of Fenton's appear to us to have the most obvious and direct bearings upon the genetic relationships of the plant furfuroids and not only *per se*. To give them their full significance we [Pg 9] must recall the later researches of Brown and Morris, which establish that cane sugar is a primary or direct product of assimilation, and that starch, which had been assumed to be a species of universal *matière première*, is probably rather a general reserve for the elaborating work of the plant. If now the aldose groups tend to pass over into the starch form, representing a temporary overflow product of the assimilating energy, it would appear that the ketose or levulose groups are preferentially used up in the elaboration of the permanent tissue. We must also take into consideration the researches of Lobry de Bruyn showing the labile functions of the typical CO group in both aldoses and hexoses, whence we may conclude that in the plant-cell the transition from dextrose to levulose is a very simple and often occurring process.

We ourselves have contributed a link in this chain of evidence connecting the furfuroids of the plant with levulose or other ketohexose. We have shown that the hydroxyfurfurals are constituents of the lignocelluloses. The proportion present in the free state is small, and it is not difficult to show that they are products of breakdown of the lignone groups. If we assume that such groups are derived ultimately from levulose, we have to account for the detachment of the methyl group. This, however, is not difficult, and we need only call to mind that the lignocelluloses are characterised by the presence of methoxy groups and a residue which is directly and easily hydrolysed to acetic acid. Moreover, the condensation need not be assumed to be a simple dehydration with attendant rearrangement; it may very well be accompanied or preceded by fixation of oxygen. Leaving out the hypothetical discussion of minor variations, there is a marked convergence of the evidence as to the main facts which establish the general relationships of the furfuroid group. This group includes both saturated and unsaturated or condensed compounds. The [Pg 10] former are constituents of celluloses, the latter of the lignone complex of the lignocelluloses.

The actual production of furfural by boiling with condensing acids is a quantitative measure of only a portion, i.e. certain members of the group. The hydroxyfurfurals, not being volatile, are not measured in this way. By secondary reactions they may yield some furfural, but as they are highly reactive compounds, and most readily condensed, they are for the most part converted into complex 'tarry' products. Hence we have no means, as yet, of estimating those tissue constituents which yield hydroxyfurfurals; also we have no measure of the furfurane-rings existing performed in such a condensed complex as lignone. But, chemists having added in the last few years a large number of facts and well-defined probabilities, it is clear that the further investigation of the furfuroid group will take its stand upon a much more adequate basis than heretofore. On the view of 'furfural-yielding' being co-extensive with 'pentose or pentosane,' not only were a number of important facts obscured or misinterpreted, but there was a barrenness of suggestion of genetic relationships. As the group has been widened very much beyond these limits, it is clear that if any group term or designation is to be retained that of 'furfuroid' is 'neutral' in character, and equally applicable to saturated substances of such widely divergent chemical character as pentoses, hexosones, glycuronic acid, and perhaps, most important of all, levulose itself, all of which are susceptible of condensation to furfural or furfurane derivatives, as well as to those unsaturated compounds, constituents of plant tissues which are already furfurane derivatives.

From the chemical point of view such terms are perhaps superfluous. But physiological relationships have a significance of their own; and there is a physiological or functional [Pg 11] cohesion marking this group which calls for recognition, at least for the time, and we therefore propose to retain the term furfuroid. [1]

General Experimental Methods. — In the investigation of the cellulose group it is clear that methods of ultimate hydrolysis are of first importance. None are so convenient as those which are based on the action of sulphuric acid, more or less concentrated

($H_2SO_4.3H_2O$ - $H_2SO_4H_2O$). Such methods have been frequently employed in the investigations noted in this volume. We notice a common deficiency in the interpretation of the results. It appears to be sufficient to isolate and identify a crystalline monose, without reference to the yield or proportion to the parent substance, to establish some main point in connection with its constitution. On the other hand, it is clear that in hydrolysing a given cellulose-complex we ought to aim at complete, i.e. *quantitative, statistics*. The hydrolytic transformation of starch to dextrins and maltose has been followed in this way, and the methods may serve as a model to which cellulose transformations should be approximated. In fact, what is very much wanted is a systematic re-examination of the typical celluloses in which all the constants of the terms between the original and the ultimate monose groups shall be determined. Such constants are similar to those for the starch-dextrose series, viz. opticity and cupric reduction. Various methods of fractionation are similarly available, chiefly the precipitation of the intermediate 'dextrins' by alcohol.

Where the original celluloses are homogeneous we should thus obtain transformation series, similarly expressed to those of starch. In the case of the celluloses which are mixtures, or of complex constitution, there are various methods of [Pg 12] either fractionating the original, or of selectively attacking particular monoses resulting from the transformation. By methods which are approximately quantitative a mixture of groups, such as we have, for instance, in jute cellulose, could be followed through the several stages of their resolution into monoses. To put the matter generally, in these colloidal and complex carbohydrates the ordinary physical criteria of molecular weight are wanting. Therefore, we cannot determine the relationship of a given product of decomposition to the parent molecule save by means of a quantitative mass-proportion. Physical criteria are only of determining value when associated with such constants as cupric reduction, and these, again, must be referred to some arbitrary initial weight, such as, for convenience, 100 parts of the original.

Instead of adopting these methods, without which, as a typical case, the mechanism of starch conversions could not have been followed, we have been content with a purely qualitative study of the analogous series obtainable from the celluloses under the action of sulphuric acid. A very important field of investigation lies open, especially to those who are generally familiar with the methods of studying starch conversions; and we may hope in this direction for a series of valuable contributions to the problem of the actual constitution of the celluloses.

FOOTNOTES:

[1] In this we are confirmed by other writers. See Tollens, *J. für Landw.* 1901, p. 27.

SECTION I. GENERAL CHEMISTRY OF THE TYPICAL COTTON CELLULOSE

(p. 3) [2] **Ash Constituents.** — It is frequently asserted that silica has a structural function *sui generis* in the plant skeleton, having a relationship to the cellulosic constituents of the plant, distinct from that of the inorganic ash components with which it is associated. It should be noted that the matter has been specifically investigated in two directions. In Berl. Ber. 5, 568 (A. Ladenburg), and again in 11, 822 (W. Lange), appear two papers 'On the Nature of Plant Constituents containing Silicon,' which contain the results of experimental investigations of equisetum species — distinguished for their exceptionally high 'ash' with large proportion of silica — to determine whether there are any grounds for assuming the existence of silicon-organic compounds in the plant, the analogues of carbon compounds. The conclusions arrived at are entirely negative. In reference to the second assumption that the cuticular tissues of cereal straws, of esparto, of the bamboo, owe their special properties to siliceous components, it has been shown by direct experiment upon the former that their rigidity and resistance to water are in no way affected by cultivation in a silica-free medium. In other words, the structural peculiarities of the gramineæ in these respects are due to the physical characteristics chiefly of the (lignified) cells of the hypodermal tissue, and to the composition and arrangement of the cells of the cuticle. [Pg 14]

'Swedish' filter papers of modern make are so far freed from inorganic constituents that the weight of the ash may be neglected in nearly all quantitative experiments [Fresenius, Ztschr. Anal Chem. 1883, 241]. It represents usually about 1/1000 mgr. per 1 sq. cm. of area of the paper.

The form of an 'ash' derived from a fibrous structure, is that of the 'organic' original, more or less, according to its proportion and composition. The proportion of 'natural ash' is seldom large enough, nor are the components of such character as to give a coherent ash, but if in the case of a fibrous structure it is combined or intimately mixed with inorganic compounds deposited within the fibres from solution, the latter may be made to yield a perfect skeleton of the fibre after burning off the organic matter. It is by such means that the mantles used in the Welsbach system of incandescent lighting are prepared. A purified cotton fabric—or yarn—is treated with a concentrated solution of the mixed nitrates of thorium and cerium, and, after drying, the cellulose is burned away. A perfect and coherent skeleton of the fabric is obtained, composed of the mixed oxides. Such mantles have fulfilled the requirements of the industry up to the present time, but later experiments forecast a notable improvement. It has been found that artificial cellulose fibres can be spun with solutions containing considerable proportions of soluble compounds of these oxides. Such fibres, when knitted into mantles and ignited, yield an inorganic skeleton of the oxides of homogeneous structure and smooth contour. De Mare in 1894, and Knofler in 1895, patented methods of preparing such cellulose threads containing the salts of thorium and cerium, by spinning a collodion containing the latter in solution. When finally ignited, after being brought into the suitable mantle form, there results a structure which proves vastly more durable than the original Welsbach mantle. The [Pg 15] cause of the superiority is thus set forth by V. H. Lewes in a recent publication (J. Soc. of Arts, 1900, p. 858): 'The alteration in physical structure has a most extraordinary effect upon the light-giving life of the mantle, and also on its strength, as after burning for a few hundred hours the constant bombardment of the mantle by dust particles drawn up by the rush of air in the chimney causes the formation of silicates on the surface of the mantle owing to silica being present in the air, and this seems to affect the Welsbach structure far more than it does the "Clamond" type, with the result that when burned continuously the Welsbach mantle falls to so low a pitch of light emissivity after 500 to 600 hours, as to be a mere shadow of its former self, giving not more than one-third of its original light, whilst the Knofler mantle keeps up its light-emitting power to a much greater extent, and the Lehner fabric is the most remarkable of all. Two Lehner mantles which have now been burning continuously in my laboratory for over 3,000 hours give at this moment a brighter light emissivity than most of the Welsbachs do in their prime.' ...'The new developments of the Clamond process form as important a step in the history of incandescent gas lighting as the discoveries which gave rise to the original mantles.'

It has further been found that the oxides themselves can be dissolved in the cellulose alkaline sulphocarbonate (viscose) solution, and artificial threads have been spun containing from 25 to 30 p.ct. of the oxides in homogeneous admixture with the cellulose. This method has obvious advantages over the collodion method both in regard to the molecular relationship of the oxides to the cellulose and to cheapness of production. [Pg 16]

UNTERSUCHUNGEN ÜBER VERSCHIEDENE BESTIMMUNGSMETHODEN DER CELLULOSE.

H. Suringar and B. Tollens (Ztschr. angew. Chem. 1896, No. 23).

INVESTIGATION OF METHODS OF DETERMINING CELLULOSE.

Introduction. — This is an exhaustive bibliography of the subject, describing also the various methods of cellulose estimation, noted in historical sequence. First, the Weende 'crude fibre' method (Henneberg) with modifications of Wattenberg, Holdefleiss, and others is dealt with. The product of this treatment, viz. 'crude fibre' is a mixture, containing furfuroids and lignone compounds. Next follows a group of processes which aim at producing a 'pure cellulose' by eliminating lignone constituents, for which the merely hydrolytic treatments of the Weende method are ineffectual. The method of F. Schulze — prolonged digestion with dilute nitric acid, with addition of chlorate — has been largely employed, though the composition of the product is more or less divergent from a 'pure cellulose.'

Dilute nitric acid at 60-80° (Cross and Bevan) and a dilute mixture of nitric and sulphuric acids (Lifschutz) have been employed for isolating cellulose from the lignocelluloses. Hoffmeister modifies the method of Schulze by substituting hydrochloric acid for the nitric acid. Treatment with the halogens associated with alkaline processes of hydrolysis is the basis of the methods of Hugo Muller (bromine water) and Cross and Bevan (chlorine gas). Lastly, the authors notice the methods based upon the action of the alkaline hydrates at high temperatures (180°) in presence of water (Lange), or of glycerin (Gabriel). The process of heating to 210° with glycerin only (Hönig) yields a very impure and ill-defined product. [Pg 17]

For comparative investigation of these processes certain celluloses and cellulosic materials were prepared as follows:

(a) *'Rag' cellulose.* — A chemical filter paper, containing only cotton and linen celluloses, was further purified by boiling with dilute acid and dilute alkali. After thorough washing it was air-dried.

(b) *Wood cellulose.* — Pine wood sawdust was treated by digestion for fourteen days with dilute nitric acid with addition of chlorate (Schulze). The mass was washed and digested with alkaline lye (1.25 p.ct. KOH), and exhaustively washed, treated with dilute acetic acid; again washed, and finally air-dried.

This product was found to yield 2.3 p.ct. furfural on distillation with HCl (1.06 sp.gr.).

(c) *Purified wood.* — Pine wood sawdust was treated in succession with dilute alkalis and acids, in the cold, and with alcohol and ether until exhausted of products soluble in these liquids and reagents.

In addition to the above the authors have also employed jute fibre and raw cotton wool in their investigations.

They note that the yield of cellulose is in many cases sensibly lowered by treating the material after drying at the temperature of 100°. The material for treatment is therefore weighed in the air-dry condition, and a similar sample weighed off for drying at 100° for determination of moisture.

The main results of the experimental investigation are as follows: —

Weende process further attacks the purified celluloses as follows: Wood cellulose losing in weight 8-9 p.ct.; filter paper, 6-7.5 p.ct., and the latter treated a second time loses a further 4-5 p.ct. It is clear, therefore, that the process is of purely empirical value. [Pg 18]

Schulze. — This process gave a yield of 47.6 p.ct. cellulose from pine wood. The celluloses themselves, treated by the process, showed losses of 1-3 p.ct. in weight, much less therefore than in the preceding case.

Hönig's method of heating with glycerin to 210° was found to yield products very far removed from 'cellulose.' The process may have a certain value in estimations of 'crude fibre,' but is dismissed from further consideration in relation to cellulose.

Lange.—The purpose of the investigation was to test the validity of the statement that the celluloses are not attacked by alkaline hydrates at 180°. Experiments with pine wood yielded a series of percentages for cellulose varying from 36 to 41; the 'purified wood' gave also variable numbers, 44 to 49 per cent. It was found possible to limit these variations by altering the conditions in the later stages of isolating the product; but further experiments on the celluloses themselves previously isolated by other processes showed that they were profoundly and variably attacked by the 'Lange' treatment, wood cellulose losing 50 per cent. of its weight, and filter paper (cellulose) losing 15 per cent. Further, a specimen of jute yielded 58 per cent. of cellulose by this method instead of the normal 78 per cent. It was also found that the celluloses isolated by the process, when subjected to a second treatment, underwent a further large conversion into soluble derivatives, and in a third treatment further losses of 5-10 per cent were obtained. The authors attach value, notwithstanding, to the process which they state to yield an 'approximately pure cellulose,' and they describe a modified method embodying the improvements in detail resulting from their investigation.

Gabriel's method of heating with a glycerin solution of alkaline hydrate is a combination of 'Hönig' and 'Lange.' [Pg 19] An extended investigation showed as in the case of the latter that the celluloses themselves are more or less profoundly attacked by the treatment—further that the celluloses isolated from lignocelluloses and other complex raw materials are much 'less pure' than those obtained by the Lange process. Thus, notably in regard to furfural yielding constituents, the latter yield 1-2 p.ct. furfural, whereas *specimens of 'jute cellulose'* obtained by the Gabriel process were found to yield *9 to 13 p.ct. furfural*.

Cross and Bevan.—Chlorination process yielded in the hands of the authors results confirming the figures given in 'Cellulose' for yield of cellulose. Investigation of the products for yield of furfural, gave 9 p.ct. of this aldehyde showing the presence of celluloses, other than the normal type.

Conclusions.—The subjoined table gives the mean numerical results for yield of end-product or 'cellulose' by the various methods. In the case of the 'celluloses' the results are those of the further action of the several processes on the end-product of a previous process.

	Methods				
	F. Schulze	Weende	Lange	Gabriel	Cross and Bevan
Wood cellulose	98.51	91.52	48.22	55.93	—
Filter paper cellulose	99.62	95.63	78.17	79.77	—
Swedish filter paper	96.58	—	84.76	—	—
Ordinary filter paper	98.17	93.39	86.58	—	—
Cotton ('wool')	98.38	89.98	63.96	67.88	—
Jute	—	—	57.93	71.64	75.27
Purified wood	—	—	{49.27 {46.56	—	—
Raw wood	47.60	—	{40.82 {38.87	—	—

The final conclusion drawn from the results is that none of the processes fulfil the requirements of an ideal method. [Pg 20] Those which may be carried out in a reasonably short time are deficient in two directions: (1) they yield a 'cellulose' containing more or less oxycellulose; (2) the celluloses themselves are attacked under the conditions of treatment, and the end product or cellulose merely represents a particular and at the same time variable equilibrium, as between the resistance of the cellulose and the attack of the reagents employed; this attack being by no means confined to the non-cellulose constituents. Schulze's method appears to give the nearest approximation to the 'actual cellulose' of the raw material.

(p. 8) **SOLUTIONS OF CELLULOSE**—(1) **ZINC CHLORIDE.**—To prepare a homogeneous solution of cellulose by means of the neutral chloride, a prolonged digestion at or about 100° with the concentrated reagent is required. The dissolution of the cellulose is not a simple phenomenon, but is attended with hydrolysis and a certain degree of condensation. The latter result is evidenced by the formation of furfural, the former by the presence of soluble carbohydrates in the solution obtained by diluting the original solution and filtering from the reprecipitated cellulose. The authors have observed that in carefully conducted experiments cotton cellulose may be dissolved in the reagent, and reprecipitated with a loss of only 1 p.ct. in weight. This, however, is a 'net' result, and leaves undetermined the degree of hydration of the recovered cellulose as of hydrolysis of the original to groups of lower molecular weights. Bronnert finds that a previous hydration of the cellulose—e.g. by the process of alkaline mercerisation and removal of the alkali by washing—enables the zinc chloride to effect its dissolution by digestion in the cold. (U.S. patent, 646,799/1900. See also p. 59.)

Industrial applications.—(a) *Vulcanised fibre* is prepared by treating paper with four times its weight of the concentrated [Pg 21] aqueous solution (65-75° B.), and in the resulting gelatinised condition is worked up into masses, blocks, sheets, &c., of any required thickness. The washing of these masses to remove the zinc salt is a very lengthy operation.

To render the product waterproof the process of nitration is sometimes superadded [D.R.P. 3181/1878]. Further details of manufacture are given in Prakt. Handbuch d. Papierfabrikation, p. 1703 [C. Hofmann].

(b) *Calico-printing.*—The use of the solution as a thickener or colour vehicle, more especially as a substitute for albumen in pigment styles, was patented by E. B. Manby, but the process has not been industrially developed [E.P. 10,466 / 1894].

(c) *Artificial silk.*—This is a refinement of the earlier applications of the solution in spinning cellulose threads for conversion into carbon filaments for electrical glow-lamps. This section will be found dealt with on p. 59.

(p. 13) (2) **Cuprammonium solution.**—The application of the solution of cellulose in cuprammonium to the production of a fine filament in continuous length, 'artificial silk,' has been very considerably studied and developed in the period 1897-1900, as evidenced by the series of patents of Fremery and Urban, Pauly, Bronnert, and others. The subject will also be found dealt with on p. 58.

(p. 15) **Reactions of cellulose with iodine.**—In a recent paper, F. Mylius deals with the reaction of starch and cellulose with iodine, pointing out that the blue colouration depends upon the presence of water and iodides. In absence of the latter, and therefore in presence of compounds which destroy or absorb hydriodic acid—e.g. iodic acid—there results a *brown* addition product. The products in question have the characteristics of *solid solutions* of the halogen. (Berl. Ber. 1895, 390.) [Pg 22]

(24) **Mercerisation**—Notwithstanding the enormous recent developments in the industrial application of the mercerising reaction, there have been no noteworthy contributions to the theoretical aspects of the subject. The following abstract gives an outline of the scope of an important technical work on the subject.

DIE MERCERISATION DER BAUMWOLLE.

Paul Gardner (Berlin: 1898. J. Springer).

THE MERCERISATION OF COTTON.

This monograph of some 150 pages is chiefly devoted to the patent literature of the subject. The chemical and physical modifications of the cotton substance under the action of strong alkaline lye, were set forth by Mercer in 1844-5, and there has resulted from subsequent investigations but little increase in our knowledge of the fundamental facts. The treatment was industrially developed by Mercer in certain directions, chiefly (1) for preparing webs of cloth required to stand considerable strain, and (2) for producing crêpon effects by local or topical action of the alkali. But the results achieved awakened but a transitory interest, and the matter passed into oblivion; so much so, indeed, that a German patent [No. 30,966] was granted in 1884 to the Messrs. Depouilly for crêpon effects due to the differential shrinkage of fabrics under mercerisation, by processes and treatments long previously described by Mercer. Such effects have had a considerable vogue in recent years, but it was not until the discovery of the lustreing effect resulting from the association of the mercerising actions with the condition of strain or tension of the yarn or fabric that the industry in 'mercerised' goods was started on the lines which have led to the present [Pg 23] colossal development. The merit of this discovery is now generally recognised as belonging to Thomas and Prevost of Crefeld, notwithstanding that priority of patent right belongs to the English technologist, H. A. Lowe.

The author critically discusses the grounds of the now celebrated patent controversy, arising out of the conflict of the claims of German patent 85,564/1895 of the former, and English patent 4452/1890 of the latter. The author concludes that Lowe's specification undoubtedly describes the lustreing effect of mercerising in much more definite terms than that of Thomas and Prevost. These inventors, on the other hand, realised the effect industrially, which Lowe certainly failed to do, as evidenced by his allowing the patent to lapse. As an explanation of his failure, the author suggests that Lowe did not sufficiently extend his observations to goods made from Egyptian and other long-stapled cottons, in which class only are the full effects of the added lustre obtained.

Following these original patents are the specifications of a number of inventions which, however, are of insignificant moment so far as introducing any essential variation of the mercerising treatment.

The third section of the work describes in detail the various mechanical devices which have been patented for carrying out the treatment on yarn and cloth.

The fourth section deals with the fundamental facts underlying the process and effects summed up in the term 'mercerisation.' These are as follows: —

(*a*) Although all forms of fibrous celluloses are similarly affected by strong alkaline solutions, it is only the Egyptian and other long-stapled cottons—i.e. the goods made from them—which under the treatment acquire the special high lustre which ranks as 'silky.' Goods made from American cottons acquire a certain 'finish' and lustre, but the effects [Pg 24] are not such as to have an industrial value—i.e. a value proportional to the cost of treatment.

(*b*) The lustre is determined by exposing the goods to strong tension, either when under the action of the alkali, or subsequently, but only when the cellulose is in the special condition of hydration which is the main chemical effect of the mercerising treatment.

(*c*) The degree of tension required is approximately that which opposes the shrinkage in dimensions, otherwise determined by the action of the alkali. The following table exhibits the variations of shrinkage of Egyptian when mercerised without tension, under varying conditions as regards the essential factors of the treatment—viz. (1) concentration of the alkaline lye, (2) temperature, and (3) duration of action (the latter being of subordinate moment):—

ation	5°B.			10°B.			15°B			25°B			30°B			35°
of	1	10	30	1	10	30	1	10	30	1	10	30	1	10	30	1
tures :—	Percentage shrinkages (Egyptian yarns) as under:—															
	0	0	0	1	1	1	12.2	15.2	15.8	19.2	19.8	21.5	22.7	22.7	22.7	24.
	0	0	0	0	0	0	8.0	8.8	11.8	19.8	20.1	21.0	21.2	22.0	22.3	23.
	0	0	0	0	0	0	4.6	4.6	6.0	19.0	19.5	19.0	18.5	19.5	19.8	20.
	0	0	0	0	0	0	3.5	3.5	9.8	13.4	13.7	14.2	15.0	15.1	15.5	15.

The more important general indications of the above results are— (1) The mercerisation action commences with a lye of 10°B., and increases with increased strength of the lye up to a maximum at

35°B. There is, however, a relatively slight increase of action with the increase of caustic soda from 30-40°B. (2) For optimum action the temperature should not exceed 15-20°C. (3) The duration of action is of proportionately less influence as the concentration of the lye increases. As the maximum effect is attained the action becomes practically instantaneous, the only condition affecting it being that of penetration—i.e. actual contact of cellulose and alkali. [Pg 25]

(*d*) The question as to whether the process of 'mercerisation' involves chemical as well as physical effects is briefly discussed. The author is of opinion that, as the degree of lustre obtained varies with the different varieties of cotton, the differentiation is occasioned by differences in chemical constitution of these various cottons. The influence of the chemical factors is also emphasised by the increased dyeing capacity of the mercerised goods, which effect, moreover, is independent of those conditions of strain or tension under mercerisation which determine lustre. It is found in effect that with a varied range of dye stuffs a given shade is produced with from 10 to 30 p.ct. less colouring matter than is required for the ordinary, i.e. unmercerised, goods.

In reference to the constants of strength and elasticity, Buntrock gives the following results of observations upon a 40^5 twofold yarn, five threads of 50 cm. length being taken for each test(Prometheus, 1897, p. 690): (*a*) the original yarn broke under a load of 1440 grms.; (*b*) after mercerisation without tension the load required was 2420 grms.; (*c*) after mercerisation under strain, 1950 grms. Mercerisation, therefore, increases the strength of the yarn from 30 to 66 p.ct., the increase being lessened proportionately to the strain accompanying mercerisation. *Elasticity*, as measured by the extension under the breaking load, remains about the same in yarns mercerised under strain, but when allowed to shrink under mercerisation there is an increase of 30-40 p.ct. over the original.

The *change of form* sustained by the individual fibres has been studied by H. Lange [Farberzeitung, 1898, 197-198], whose microphotographs of the cotton fibres, both in length and cross-section, are reproduced. In general terms, the change is from the flattened riband of the original fibre to a cylindrical tube with much diminished and rounded central canal. The effect of strain under mercerisation is chiefly seen [Pg 26] in the contour of the surface, which is smooth, and the obliteration at intervals of the canal. Hence the increased transparency and more complete reflection of the light from the surface, and the consequent approximation to the optical properties of the silk fibre.

The work concludes with a section devoted to a description of the various practical systems of mercerisation of yarns in general practice in Germany, and an account of the methods adopted in dyeing the mercerised yarns.

RESEARCHES ON MERCERISED COTTON.

A. Fraenkel and P. Friedlaender (Mitt. k.-k. Techn. Gew. Mus., Wien, 1898, 326).

The authors, after investigation, are inclined to attribute the lustre of mercerised cotton to the absence of the cuticle, which is destroyed and removed in the process, partly by the chemical action of the alkali, and partly by the stretching at one or other stage of the process. The authors have investigated the action of alcoholic solutions of soda also. The lustre effects are not obtained unless the action of water is associated.

In conclusion, the authors give the following particulars of breaking strains and elasticity: —

Treatment	Experiments	Breaking strain Grammes	Elasticity Elongation in mm.
Cotton unmercerised.	1	360	20
	2	356	20
	3	360	22
Mercerised with Soda 35°B.	1	530	44
	2	570	40

	3	559	35
Alcoholic soda 10 p.ct. cold	1	645	24
	2	600	27
	3	610	33
Alcoholic soda 10 p.ct. hot	5	740	33
	2	730	38
	3	690	30

FOOTNOTES:

[2] This and other similar references are to the matter of the original volume (1895).

[Pg 27]

SECTION II. SYNTHETICAL DERIVATIVES — SULPHOCARBONATES AND ESTERS

(p. 25) **Cellulose sulphocarbonate.** — Further investigations of the reaction of formation as well as the various reactions of decomposition of the compound, have not contributed any essential modification or development of the subject as originally described in the author's first communications. A large amount of experimental matter has been accumulated in view of the ultimate contribution of the results to the general theory of colloidal solutions. But viscose is a complex product and essentially variable, through its pronounced tendency to progressive decomposition with reversion of the cellulose to its insoluble and uncombined condition. The solution for this reason does not lend itself to exact measurement of its physical constants such as might elucidate in some measure the progressive molecular aggregation of the cellulose in assuming spontaneously the solid (hydrate) form. Reserving the discussion of these points, therefore, we confine ourselves to recording results which further elucidate special points.

Normal and other celluloses. — We may certainly use the sulphocarbonate reaction as a means of defining a normal cellulose. As already pointed out, cotton cellulose passes quantitatively through the cycle of treatments involved in solution as sulphocarbonate and decomposition of the solution with regeneration as structureless or amorphous cellulose (hydrate).

Analysis of this cellulose shows a fall of carbon percentage from 44.4 to 43.3, corresponding with a change in composition from $C_6H_{10}O_5$ to $4C_6H_{10}O_5.H_2O$. The partial hydrolysis affects the whole molecule, and is limited to this effect, whereas, in the case of celluloses of other types, there is a fractionation of the mass, a portion undergoing a further hydrolysis to compounds of lower molecular weight and permanently soluble. Thus in the case of the wood celluloses [Pg 28] the percentage recovered from solution as viscose is from 93 to 95 p.ct. It is evident that these celluloses are not homogeneous. A similar conclusion results from the presence of furfural-yielding compounds with the observation that the hydrolysis to soluble derivatives mainly affects these derivatives. In the empirical characterisation of a normal cellulose, therefore, we may include the property of quantitative regeneration or recovery from its solution as sulphocarbonate.

In the use of the word 'normal' as applied to a 'bleached' cotton, we have further to show in what respects the sulphocarbonate reaction differentiates the bleached or purified cotton cellulose from the raw product. The following experiments may be cited: Specimens of American and Egyptian cottons in the raw state, freed from mechanical, i.e. non-fibrous, impurities, were treated with a mercerising alkali, and the alkali-cotton subsequently exposed to carbon disulphide. The product of reaction was further treated as in the preparation of the ordinary solution; but in place of the usual solution, structureless and homogeneous, it was observed to retain a fibrous character, and the fibres, though enormously swollen, were not broken down by continued vigorous stirring. After large dilution the solutions were filtered, and the fibres then formed a gelatinous mass on the filters. After purification, the residue was dried and weighed. The American cotton yielded 90.0 p.ct., and the Egyptian 92.0 p.ct. of its substance in the form of this peculiar modification. The experiment was repeated, allowing an interval of 24 hours to elapse between the conversion into alkali-cotton and exposure of this to the carbon disulphide. The quantitative results were identical.

There are many observations incidental to chemical treatments of cotton fabrics which tend to show that the bleaching process produces other effects than the mere [Pg 29] removal of mechanical impurities. In the sulphocarbonate reaction the raw cotton, in fact, behaves exactly as a compound cellulose. Whether the constitutional difference between raw and bleached cotton, thus emphasised, is due to the group of components of the raw cotton, which are removed in the bleaching process, or to internal constitutional changes determined by the bleaching treatments, is a question which future investigation must decide.

The normal sulphocarbonate (viscose). — In the industrial applications of viscose it is important to maintain a certain standard of composition as of the essential physical properties of the solution, notably viscosity. It may be noted first that, with the above-mentioned exception, the various fibrous celluloses show but slight differences in regard to all the essential features of the reactions involved. In the mercerising reaction, or alkali-cellulose stage, it is true the differences are considerable. With celluloses of the wood and straw classes there is a considerable conversion into soluble alkali-celluloses. If treated with water these are dissolved, and on weighing back the cellulose, after thorough washing, treatment with acid, and finally washing and drying, it will be found to have lost from 15 to 20 p.ct. in weight. The lower grade of celluloses thus dissolved are only in part precipitated in acidifying the alkaline solution. On the other hand, after conversion into viscose, the cellulose when regenerated re-aggregates a large proportion of these lower grade celluloses, and the final loss is as stated above, from 5 to 7 p.ct. only.

Secondly, it is found that all the conditions obtaining in the alkali-cellulose stage affect the subsequent viscose reaction and the properties of the final solution. The most important are obviously the proportion of alkali to cellulose and the length of time they are in contact before being treated with carbon disulphide. An excess of alkali beyond the 'normal' [Pg 30] proportion—viz. 2NaOH per 1 mol. $C_6H_{10}O_5$—has little influence upon the viscose reaction, but lowers the viscosity of the solution of the sulphocarbonate prepared from it. But this effect equally follows from addition of alkali to the viscose itself. The alkali-cellulose changes with age; there is a gradual alteration of the molecular structure of the cellulose, of which the properties of the viscose when prepared are the best indication. There is a progressive loss of viscosity of the solution, and a corresponding deterioration in the structural properties of the cellulose when regenerated from it—especially marked in the film form. In regard to viscosity the following observations are typical:—

(*a*) A viscose of 1.8 p.ct. cellulose prepared from an alkali-cellulose (cotton) fourteen days old.

(*b*) Viscose of 1.8 p.ct. cellulose from an alkali-cellulose (cotton) three days old.

(*c*) Glycerin diluted with 1/3 vol. water.

	a	*b*	*b*	*c*
			Diluted with	

		equal vol. water	
Times of flow of equal volumes from narrow orifice in seconds	112 321	103	170

Similarly the cellulose in reverting to the solid form from these 'degraded' solutions presents a proportionate loss of cohesion and aggregating power expressed by the inferior strength and elasticity of the products. Hence, in the practical applications of the product where the latter properties are of first importance, it is necessary to adopt normal standards, such as above indicated, and to carefully regulate all the conditions of treatment in each of the two main stages of reaction, so that a product of any desired character may be invariably obtained.

Incidentally to these investigations a number of observations have been made on the alkali-cellulose (cotton) after [Pg 31] prolonged storage in closed vessels. It is well known that starch undergoes hydrolysis in contact with aqueous alkalis of a similar character to that determined by acids [Béchamp, Annalen, 100, 365]. The recent researches of Lobry de Bruyn [Rec. Trav. Chim. 14, 156] upon the action of alkaline hydrates in aqueous solution on the hexoses have established the important fact of the resulting mobility of the CO group, and the interchangeable relationships of typical aldoses and ketoses. It was, therefore, not improbable that profound hydrolytic changes should occur in the cellulose molecule when kept for prolonged periods as alkali-cellulose.

We may cite an extreme case. A series of products were examined after 12-18 months' storage. They were found to contain only 3-5 p.ct. 'soluble carbohydrates'; these were precipitated by Fehling's solution but without reduction on boiling. They were, therefore, of the cellulose type. On acidifying with sulphuric acid and distilling, traces only of volatile acid were produced. It is clear, therefore, that the change of molecular weight of the cellulose, the disaggregation of the undoubtedly large molecule of the original 'normal' cellulose—which effects are immediately recognised in the viscose reactions of such products—are of such otherwise limited character that they do not affect the constitution of the unit groups. We should also conclude that the cellulose type of constitution covers a very wide range of minor variations of molecular weight or aggregation.

The resistance of the normal cellulose to the action of alkalis under these hydrolysing conditions should be mentioned in conjunction with the observations of Lange, and the results of the later investigations of Tollens, on its resistance to 'fusion' with alkaline hydrates at high temperatures (180°). The degree of resistance has been established only on the empirical basis of weighing the product recovered from such [Pg 32] treatment. The product must be investigated by conversion into typical cellulose derivatives before we can pronounce upon the constitutional changes which certainly occur in the process. But for the purpose of this discussion it is sufficient to emphasise the extraordinary resistance of the normal cellulose to the action of alkalis, and to another of the more significant points of differentiation from starch.

Chemical constants of cellulose sulphocarbonate (solution). — In investigations of the solutions we make use of various analytical methods, which may be briefly described, noting any results bearing upon special points.

Total alkali. — This constant is determined by titration in the usual way. The cellulose ratio, $C_6H_{10}O_5$: 2NaOH, is within the ordinary error of observation, 2:1 by weight. A determination of alkali therefore determines the percentage of cellulose.

Cellulose may be regenerated in various ways — viz. by the action of heat, of acids, of various oxidising compounds. It is purified for weighing by boiling in neutral sulphite of soda (2 p.ct. solution) to remove sulphur, and in very dilute acids (0.33 p.ct. HCl) to decompose residues of 'organic' sulphur compounds. It may also be treated with dilute oxidants. After weighing it may be ignited to determine residual inorganic compounds.

Sulphur. — It has been proved by Lindemann and Motten [Bull. Acad. R. Belg. (3), 23, 827] that the sulphur of sulphocarbonates (as well as of sulphocyanides) is fully oxidised (to SO_3) by the hypochlorites (solutions at ordinary temperatures). The method may be adapted as required for any form of the products or by-products of the viscose reaction to be analysed for *total sulphur*.

The sulphur present in the form of dithiocarbonates, including the typical cellulose xanthogenic acid, is approximately [Pg 33] isolated and determined as CS_2 by adding a zinc salt in excess, and distilling off the carbon disulphide from a water bath. From freshly prepared solutions a large proportion of the disulphide originally interacting with the alkali and cellulose is recovered, the result establishing the general conformity of the reaction to that typical of the alcohols. On keeping the solutions there is a progressive interaction of the bisulphide and alkali, with formation of trithiocarbonates and various sulphides. In decomposing these products by acid reagents hydrogen sulphide and free sulphur are formed, the estimation of which presents no special difficulties.

In the spontaneous decomposition of the solution a large proportion of the sulphur resumes the form of the volatile disulphide. This is approximately measured by the loss in total sulphur in the following series of determinations, in which a viscose of 8.5 p.ct. strength (cellulose) was dried down as a thin film upon glass plates, and afterwards analysed:

(*a*) Proportion of sulphur to cellulose (100 pts.) in original.
(*b*) After spontaneous drying at ordinary temperature.
(*c*) After drying at 40°C.
(*d*) As in (*c*), followed, by 2 hours' heating at 98°.
(*e*) As in (*c*), followed by 5 hours' heating at 98°.

	a	b	c	d	e
Total sulphur	40.0	25.0	31.0	23.7	10.4

The dried product in (*b*) and (*c*) was entirely resoluble in water; in (*d*) and (*e*), on the other hand, the cellulose was fully regenerated, and obtained as a transparent film.

Iodine reaction. — Fresh solutions of the sulphocarbonate show a fairly constant reaction with normal iodine solution. At the first point, where the excess of iodine visibly persists, there is complete precipitation of the cellulose as the bixanthic sulphide; and this occurs when the proportion of iodine added reaches $3I_2 : 4Na_2O$, calculated to the total alkali. [Pg 34]

Other decompositions.—The most interesting is the interaction which occurs between the cellulose xanthogenate and salts of ammonia, which is taken advantage of by C. H. Stearn in his patent process of spinning artificial threads from viscose. The insoluble product which is formed in excess of the solution of ammonia salt is free from soda, and contains 9-10 p.ct. total sulphur. The product retains its solubility in water for a short period. The solution may be regarded as containing the ammonium cellulose xanthate. This rapidly decomposes with liberation of ammonia and carbon disulphide, and separation of cellulose (hydrate). As precipitated by ammonium-chloride solution the gelatinous thread contains 15 p.ct. of cellulose, with a sp.gr. 1.1. The process of 'fixing'—i.e. decomposing the xanthic residue—consists in a short exposure to the boiling saline solution. The further dehydration, with increase of gravity and cellulose content, is not considerable. The thread in its final air-dry state has a sp.gr. 1.48.

Cellulose Benzoates.—These derivatives have been further studied by the authors. The conditions for the formation of the mono-benzoate [$C_6H_9O_4.O.CO.Ph$] are very similar to those required for the sulphocarbonate reaction. The fibrous cellulose (cotton), treated with a 10 p.ct. solution NaOH, and subsequently with benzoyl chloride, gives about 50 p.ct. of the theoretical yield of monobenzoate. Converted by 20 p.ct. solution NaOH into alkali-cellulose, and with molecular proportions as below, the following yields were obtained:—

		Calc. for Monobenzoate
(a) $C_6H_{10}O_5$: 2.0-2.5 NaOH : $C_6H_5.COCl$ —	150.8}	
(b) $C_6H_{10}O_5$: 2.0-2.5 NaOH : 1.5 mol. C_6H_5COCl	159.0}	164.0

An examination of (a) showed that some dibenzoate (about 7 p.ct.) had been formed. The product () was exhaustively [Pg 35] treated with cuprammonium solution, to which it yielded about 20 p.ct. of its weight, which was therefore unattacked cellulose.

Under conditions as above, but with 2.5 mol. C_6H_5COCl, a careful comparison was made of the behaviour of the three varieties of cotton, which were taken in the unspun condition and previously fully bleached and purified.

	Sea Island	Egyptian	American
Aggregate yield of benzoate	153	148	152
Moisture in air dry state	5.28	5.35	5.15
Proportion of dibenzoate p.ct.	8.30	13.70	9.4
Yield of cellulose by saponification	58.0	54.0	58.3

It appears from these results that the benzoate reaction may proceed to a higher limit (dibenzoate) in the case of Egyptian cotton. This would necessarily imply a higher limit of 'mercerisation,' under equal conditions of treatment with the alkaline hydrate. It must be noted that in the conversion of the fibrous cellulose into these (still) fibrous monobenzoates, there are certain mechanical conditions imported by the structural features of the ultimate fibres. For the elimination of the influence of this factor a large number of quantitative comparisons will be necessary. The above results are therefore only cited as typical of a method of comparative investigation, more especially of the still open questions of the cause of the superior effects in mercerisation of certain cottons (see p. 23). It is quite probable that chemical as well as structural factors co-operate in further differentiating the cottons.

Further investigation of the influence upon the benzoate reaction, of increase of concentration of the soda lye, used in the preliminary alkali cellulose reaction, from 20 to 33 p.ct. NaOH, established (1) that there is no corresponding increase in the benzoylation, and (2) that this ester reaction and the [Pg 36] sulphocarbonate reaction are closely parallel, in that the degree and limit of reaction are predetermined by the conditions of formation of the alkali cellulose.

Monobenzoate prepared as above described is resistant to all solvents of cellulose and of the cellulose esters, and is therefore freed from cellulose by treatment with the former, and from the higher benzoate by treatment with the latter. Several of these, notably pyridine, phenol and nitrobenzene, cause considerable swelling and gelatinisation of the fibres, but without solution.

Structureless celluloses of the 'normal' type, and insoluble therefore in alkaline lye, treated under similar conditions to those described above for the fibrous celluloses, yield a higher proportion of dibenzoate. The following determinations were made with the cellulose (hydrate) regenerated from the sulphocarbonate:—

Mol. proportions of reagents	Yield	Dibenzoate p.ct.
$C_6H_{10}O_5 : 2NaOH : 2BzCl$	145	34.7
[Caustic soda at 10 per cent. NaOH]		
$C_6H_{10}O_5 : 4NaOH : 2BzCl$	162	62.7
[Caustic soda at 20 per cent. NaOH]		

Limit of reaction.—The cellulose in this form having shown itself more reactive, it was taken as the basis for determining the maximum proportion of OH groups yielding to this later reaction. The systematic investigations of Skraup [Monatsh. 10, 389] have determined that as regards the interacting groups the molecular proportions 1 OH: 7 NaOH: 5 BzCl, ensure complete or maximum esterification. The maximum of OH groups in cellulose being 4, the reagents were taken in the proportion $C_6H_{10}O_5$: 4 [7 NaOH: 5 BzCl]. The yield of crude product, after purifying as far as possible from the excess of benzoic acid, was 240 p.ct. [calculated for dibenzoate 227 p.ct.]. On further investigating the crude product by treatment with solvents, it was found to have still retained benzoic acid. [Pg 37] There was also present a proportion of only partially attacked cellulose (monobenzoate). The soluble benzoate amounted to 90 p.ct. of the product. It may be generally concluded that the dibenzoate represents the normal maximum but that with the hydrated and partly hydrolysed cellulose molecule, as obtained by regeneration from the sulphocarbonate, other OH groups may react, but they are only a fractional proportion in relation to the unit group $C_6H_{10}O_5$. In this respect again there is a close parallelism between the sulphocarbonate and benzoyl-ester reactions.

The dibenzoate, even when prepared from the fibrous celluloses, is devoid of structure, and its presence in admixture with the fibrous monobenzoate is at once recognised as it constitutes a structureless incrustation. Under the microscope its presence in however minute proportion is readily observed. As stated it is soluble in certain of the ordinary solvents of the cellulose esters, e.g. chloroform, acetic acid, nitrobenzene, pyridine, and phenol. It is not soluble in ether or alcohol.

Hygroscopic moisture of benzoates.—The crude monobenzoate retains 5.0-5.5 p.ct. moisture in the air-dry condition. After removal of the residual cellulose this is reduced to 3.3 p.ct. under ordinary atmospheric conditions. The purified dibenzoates retain 1.6 p.ct. under similar conditions.

Analysis of benzoates.—On saponification of these esters with alcoholic sodium hydrate, anomalous results are obtained. The acid numbers, determined by titration in the usual way, are 10-20 p.ct. in excess of the theoretical, the difference increasing with the time of boiling. Similarly the residual cellulose shows a deficiency of 5-9 p.ct.

It is by no means improbable that in the original ester reaction there is a constitutional change in the cellulose molecule causing it to break down in part under the hydrolysing treatment with formation of acid products. This point is under [Pg 38] investigation. Normal results as regards acid numbers, on the other hand, are obtained by saponification with sodium ethylate in the cold, the product being digested with the half-saturated solution for 12 hours in a closed flask.

The following results with specimens of mono- and dibenzoate, purified, as far as possible, may be cited:

		Combustion results		Saponification results			
			Calc.	$C_6H_5.COOH$	Calc.	Cellulose	Calc.
Monobenzoate	C	56.60	58.65}	46.0	45.9	58.0	60.8
	H	5.06	5.26}				
Dibenzoate	C	63.10	64.86}	65.5	66.6	34.3	40.3
	H	3.40	4.86}				

The divergence of the numbers, especially for the dibenzoate, in the case of the hydrogen, and yield of cellulose on hydrolysis are noteworthy. They confirm the probability of the occurrence of secondary changes in the ester reactions.

Action of nitrating acid upon the benzoates. — From the benzoates above described, mixed nitro-nitric esters are obtained by the action of the mixture of nitric and sulphuric acids. The residual OH groups of the cellulose are esterified and substitution by an NO_2 group takes place in the aromatic residue, giving a mixed nitric nitrobenzoic ester. The analysis of the products points to the entrance of 1 NO_2 group in the benzoyl residue in either case; in the cellulose residue 1 OH readily reacts. Higher degrees of nitration are attained by the process of solution in concentrated nitric acid and precipitation by pouring into sulphuric acid. In describing these mixed esters we shall find it necessary to adopt the C_{12} unit formula.

In analysing these products we have employed the Dumas method for *total nitrogen*. For the $O.NO_2$ groups we have found the nitrometer and the Schloesing methods to give concordant results. For the NO_2 groups it was thought that Limpricht's method, based upon reduction with stannous chloride in acid [Pg 39] solution (HCl), would be available. The quantitative results, however, were only approximate, owing to the difficulty of confining the reduction to the NO_2 groups of the nitrobenzoyl residue. By reduction with ammonium sulphide the $O.NO_2$ groups were entirely removed as in the case of the cellulose nitrates; the NO_2 was reduced to NH_2 and there resulted a cellulose amidobenzoate, which was diazotised and combined with amines and phenols to form yellow and red colouring matters, the reacting residue remaining more or less firmly combined with the cellulose.

Cellulose dinitrate-dinitrobenzoate, and cellulose trinitrate-dinitrobenzoate. — On treating the fibrous benzoate — which is a dibenzoate on the C_{12} basis — with the acid mixture under the usual conditions, a yellowish product is obtained, with a yield of 140-142 p.ct. The nitrobenzoate is insoluble in ether alcohol, but is soluble in acetone, acetic acid, and nitrobenzene. In purifying the product the former solvent is used to remove any cellulose nitrates. To obtain the maximum combination with nitroxy-groups, the product was dissolved in concentrated nitric acid, and the solution poured into sulphuric acid.

The following analytical results were obtained (a) for the product obtained directly from the fibrous benzoate and purified as indicated, (b) for the product from the further treatment of (a) as described:

		Found		Calc. for	
		(a)	(b)	Dinitrate dinitrobenzoate	Trinitrate dinitrobenzoate
Total	Nitrogen	7.84	8.97	7.99	9.24
$O.NO_2$	"	5.00	5.45	4.00	5.54
NO_2	" (Aromatic)	2.84	3.52	3.99	3.70

With the two benzoyl groups converted into nitro-benzoyl in each product, the limit of the ester reaction with the cellulose residue is reached at the third OH group. [Pg 40]

The nitrogen in the amidobenzoate resulting from the reduction with ammonium sulphide was 4.5 p.ct. — as against 5.0 p.ct. calculated. The moisture retained by the fibrous nitrate — nitrobenzoate — in the air-dry state was found to be 1.97 p.ct.

The product from the structureless dibenzoate or tetrabenzoate on the C_{12} formula, was prepared and analysed with the following results:

			Calc. for Mononitrate tetranitrobenzoate
Total	Nitrogen	6.76	7.25
$O.NO_2$	"	1.30	1.45
NO_2	" (Aromatic)	5.46	5.80

The results were confirmed by the yield of product, viz. 131 p.ct. as against the calculated 136 p.ct. They afford further evidence of the generally low limit of esterification of the cellulose molecule. From the formation of a 'normal' tetracetate — i.e. octacetate of the

C_{12} unit — we conclude that 4/5 of the oxygen atoms are hydroxyl oxygen. Of the 8 OH groups five only react in the mixed esters described above, and six only in the case of the simple nitric esters. The ester reactions are probably not simple, but accompanied by secondary reactions within the cellulose molecule.

(p. 34) **Cellulose Acetates.**—In the first edition (p. 35) we have committed ourselves to the statement that 'on boiling cotton with acetic anhydride and sodium acetate no reaction occurs.' This is erroneous. The error arises, however, from the somewhat vague statements of Schutzenberger's researches which are current in the text-books [e.g. Beilstein, 1 ed. p. 586] together with the statement that reaction only occurs at elevated temperatures (180°). As a matter of fact, reaction takes place at the boiling temperature of the anhydride. [Pg 41] We have obtained the following results with bleached cotton:

		Yield	Calc. for Monoacetate $C_6H_7O_4O.C_2H_3O$
Ester reaction		121 p.ct.	125 p.ct.
Saponification	{Cellulose	79.9	79.9
	{Acetic acid	29.9	29.4

This product is formed without apparent structural alteration of the fibre. It is entirely insoluble in all the ordinary solvents of the higher acetates. Moreover, it entirely resists the actions of the special solvents of cellulose — e.g. zinc chloride and cuprammonium. The compound is in other respects equally stable and inert. The hygroscopic moisture under ordinary atmospheric conditions is 3.2 p.ct.

Tetracetate.—This product is now made on the manufacturing scale: it has yet to establish its industrial value.

NITRIRUNG VON KOHLENHYDRATEN.

W. Will und P. Lenze (Berl. Ber., 1898, 68).

NITRATES OF CARBOHYDRATES.

(p. 38) The authors have studied the nitric esters of a typical series of the now well-defined carbohydrates—pentoses, hexoses, both aldoses and ketoses—bioses and trioses, the nitrates being prepared under conditions designed to produce the highest degree of esterification. Starch, wood, gum, and cellulose were also included in the investigations. The products were analysed and their physical properties determined. They were more especially investigated in regard to temperatures of decomposition, which were found to lie considerably lower than that of the cellulose nitrates. They also show marked and variable instability at 50° C. A main purpose of the inquiry was to throw light upon a probable cause of the instability of the cellulose nitrates, viz. the presence of nitrates [Pg 42] of hydrolysed products or carbohydrates of lower molecular weight.

The most important results are these:

Monoses.—The *aldoses* are fully esterified, in the pentoses 4 OH, in the hexoses 5 OH groups reacting. The pentose nitrates are comparatively stable at 50°; the hexose nitrates on the other hand are extremely unstable, showing a loss of weight of 30-40 p.ct. when kept 24 hours at this temperature.

Xylose is differentiated by tending to pass into an anhydride form ($C_5H_{10}O_5$-H_2O) under this esterification. When treated in fact with the mixed acids, instead of by the process usually adopted by the authors of solution in nitric acid and subsequent addition of the sulphuric acid, it is converted into the dinitrate $C_5H_6O_2.(NO_3)_2$.

Ketoses (C_6).—These are sharply differentiated from the corresponding aldoses by giving *tri*nitrates $C_6H_7O_2(NO_3)_3$ instead of *penta*nitrates, the remaining OH groups probably undergoing internal condensation. The products are, moreover, *extremely stable*. It is also noteworthy that levulose gave this same product, the trinitrate of the anhydride (levulosan) by both methods of nitration (*supra*).

The bisaccharides or bioses all give the octonitrates. The degree of instability is variable. Cane-sugar gives a very unstable nitrate. The lactose nitrate is more stable. Thus at 50° it loses only 0.7 p.ct. in weight in eight days; at 75° it loses 1 p.ct. in twenty-four hours, but with a rapid increase to 23 p.ct. in fifty-four hours. The maltose octonitrate melts (with decomposition) at a relatively high temperature, 163°-164°. At 50°-75° it behaves much like the lactose nitrate.

Trisaccharide. — Raffinose yielded the product

$C_{18}H_{21}O_5.(NO_3)_{11}$.

Starch yields the hexanitrate (C_{12}) by both methods of nitration. The product has a high melting and decomposing point, [Pg 43] viz. 184°, and when thoroughly purified is quite stable. It is noted that a yield of 157 p.ct. of this nitrate was obtained, and under identical conditions cellulose yielded 170 p.ct.

Wood gum, from beech wood, gave a tetranitrate (C_{10} formula) insoluble in all the usual solvents for this group of esters.

The authors point out in conclusion that the conditions of instability and decomposition of the nitrates of the monose-triose series are exactly those noted with the cellulose nitrates as directly prepared and freed from residues of the nitrating acids. They also lay stress upon the superior stability of the nitrates of the anhydrides, especially of the ketoses.

NITRATED CARBOHYDRATES AS FOOD MATERIAL FOR MOULDS.

Thomas Bokorny (Chem. Zeit., 1896, 20, 985-986).

(p. 38) Cellulose trinitrate (nitrocellulose) will serve as a food supply for moulds when suspended in distilled water containing the requisite mineral matter and placed in the dark. The growth is rapid, and a considerable quantity of the vegetable growth accumulates round the masses of cellulose nitrate, but no growth is observed if mineral matter is absent. Cellulose itself cannot act as a food supply, and it seems probable that if glycerol is present cellulose nitrate is no longer made use of.

NITRATION OF CELLULOSE, HYDROCELLULOSE, AND OXYCELLULOSE.

Leo Vignon (Compt. rend., 1898, 126, 1658-1661).

(p. 38) Repeated treatment of cellulose, hydrocellulose, and oxycellulose with a mixture of sulphuric and nitric acids in [Pg 44] large excess, together with successive analyses of the compounds produced, showed that the final product of the reaction corresponded, in each case, with the fixation of 11 NO groups by a molecule containing 24 atoms of carbon. On exposure to air, nitrohydrocellulose becomes yellow and decomposes; nitro-oxycellulose is rather more stable, whilst nitrocellulose is unaffected. The behaviour of these nitro-derivatives with Schiff's reagent, Fehling's solution, and potash show that all three possess aldehydic characters, which are most marked in the case of nitro-oxycellulose. The latter also, when distilled with hydrochloric acid, yields a larger proportion of furfuraldehyde than is obtained from nitrocellulose and nitrohydrocellulose.

CELLULOSE NITRATES-EXPLOSIVES.

(p. 38) The uses of the cellulose nitrates as a basis for explosives are limited by their fibrous character. The conversion of these products into the structureless homogeneous solid or semi-solid form has the effect of controlling their combustion. The use of nitroglycerin as an agent for this purpose gives the curious result of the admixture of two high or blasting explosives to produce a new explosive capable of extended use for military purposes. The leading representatives of this class of propulsive explosives, or 'smokeless powders' are ballistite and cordite, the technology of which will be found fully discussed in special manuals of the subject. Since the contribution of these inventions to the development of cellulose chemistry does not go beyond the broad, general facts above mentioned, we must refer the reader for technical details to the manuals in question.

There are, however, other means of arriving at structureless cellulose nitrates. One of these has been recently disclosed, and as the results involve chemical and technical points of [Pg 45] novelty, which are dealt with in a scientific communication, we reproduce the paper in question, viz.: —

A RE-INVESTIGATION OF THE CELLULOSE NITRATES.

A. Luck and C. F. Cross (J. Soc. Chem. Ind., 1900).

The starting-point of these investigations was a study of the nitrates obtained from the structureless cellulose obtained from the sulphocarbonate (viscose). This cellulose in the form of a fine meal was treated under identical conditions with a sample of pure cotton cellulose, viz. digested for 24 hours in an acid mixture containing in 100 parts $HNO_3 - 24 : H_2SO_4 - 70: H_2O - 6$: the proportion of acid to cellulose being 60 : 1 —. After careful purification the products were analysed with the following results:

	Nitrogen	Soluble in Ether alcohol
Fibrous nitrate	13.31	4.3 p.ct.
Structureless nitrate	13.35	5.6 "

Examined by the 'heat test' (at 80°) and the 'stability test' (at 135°) they exhibited the usual instability, and in equal degrees. Nor were the tests affected by exhaustive treatment with ether, benzene, and alcohol. From this it appears that the process of solution as sulphocarbonate and regeneration of the cellulose, though it eliminates certain constituents of an ordinary bleached cellulose, which might be expected to cause instability, has really no effect in this direction. It also appears that instability may be due to by-products of the esterification process derived from the cellulose itself.

The investigation was then extended to liquids having a direct solvent action on these higher nitrates, more especially acetone. It was necessary, however, to avoid this solvent [Pg 46] action proper, and having observed that dilution with water in increasing proportions produced a graduated succession of physical changes in the fibrous ester, we carried out a series of treatments with such diluted acetones. Quantities of the sample (A), purified as described, but still unstable, were treated each with five successive changes of the particular liquid, afterwards carefully freed from the acetone and dried at 40°C. The products, which were found to be more or less disintegrated, were then tested by the ordinary heat test, stability test, and explosion test, with the results shown in the table on next page.

In this series of trials the sample 'A' was used in the condition of pulp, viz. as reduced by the process of wet-beating in a Hollander. A similar series was carried out with the guncotton in the condition in which it was directly obtained from the ester reaction. The results were similar to above, fully confirming the progressive character of the stabilisation with increasing proportions of acetone. These results prove that washing with the diluted acetone not only rendered the nitrate perfectly stable, but that the product was more stable than that obtained by the ordinary process of purification, viz. long-continued boiling and washing in water. We shall revert to this point after briefly dealing with the associated phenomenon of structural disintegration. This begins to be well marked when the proportion of acetone exceeds 80 p.ct. The optimum effect is obtained with mixtures of 90 to 93 acetone and 10 to 7 water (by volume). In a slightly diluted acetone of such composition, the guncotton is instantly attacked, the action being quite different from the gelatinisation which precedes solution in the undiluted solvent. The fibrous character disappears, and the product assumes the form of a free, bulky, still opaque mass, which rapidly sinks to the bottom of the containing vessel. The disintegration of the bulk of the nitrate is associated with [Pg 47]

	Proportions by volume				
--	Acetone	Water	Temperature of Explosion	Heat Test 80°	Heat Test 134°
From 'A'			Deg.	Mins.	Mins.

sample.	20	80	137	3	4
	30	70	160	3	4
	40	60	180	7	18
					No fumes after
	50	50	187.5	55	100
	60	40	187	45	100
	70	30	185	45	100
	80	20	--	50	100
	92	8	185	50	100
	Structureless powder.				
" 'B' sample	50	50	183	35	100
" 'C' sample	Ordinary service guncotton		185	10	41

a certain solvent action, and on adding an equal bulk of water, the dissolved nitrate for the most part is precipitated, at the same time that the undissolved but disintegrated and swollen product undergoes further changes in the direction of increase of hardness and density. The product being now collected on a filter, freed from acetone by washing with water and dried, is a hard and dense powder the fineness of which varies according to the attendant conditions of treatment. With the main product in certain cases there is found associated a small proportion of nitrate retaining a fibrous character, which may be separated by means of a fine sieve. On examining such a residue, we found it to contain only 5.6 p.ct. N, and as it was insoluble in strong acetone, it may be regarded as a low nitrate or a mixture of such with unaltered cellulose. Confirming this we found that the product passing through the sieve showed an increase of nitrogen to 13.43 p.ct. from the 13.31 p.ct. in the original. Tested by the heat test (50 minutes) and stability test (no fumes after 100 minutes), we found the products to have the characteristics previously noticed. [Pg 48]

It is clear, therefore, that this specifically regulated action of acetone produces the effects (*a*) of disintegration, and (*b*) stabilisation. It remains to determine whether the latter effect was due, as might be supposed, to the actual elimination of a compound or group of compounds present in the original nitrate, and to be regarded as the effective cause of instability. It is to be noted first that as a result of the treatment with the diluted acetone and further dilution after the specific action is completed, collecting the disintegrated product on a filter and washing with water, the loss of weight sustained amounts to 3 to 4 p.ct. This loss is due, therefore, to products remaining dissolved in the filtrate — that is to say, in the much diluted acetone. These filtrates are in fact opalescent from the presence of a portion of nitrate in a colloidal (hydrated) form. On distilling off the acetone, a precipitation is determined. The precipitates are nitrates of variable composition, analysis showing from 9 to 12 p.ct. of nitric nitrogen. The filtrate from these precipitates containing only fractional residues of acetone still shows opalescence. On long-continued boiling a further precipitation is determined, the filtrates from which are clear. It was in this final clear filtrate that the product assumed to cause the instability of the original nitrate would be present. The quantity, however, is relatively so small that we have only been able to obtain and examine it as residue from evaporation to dryness. An exhaustive qualitative examination established a number of negative characteristics, with the conclusion that the products were not direct derivatives of carbohydrates nor aromatic compounds. On the other hand the following positive points resulted. Although the original diluted acetone extract was neutral to test papers, yet the residue was acid in character. It contained combined nitric groups, fused below 200° giving off acid vapours, and afterwards burning with a smoky flame. On adding lead acetate to the original clear solution, a well-marked [Pg 49] precipitation was determined. The lead compounds thus isolated are characteristic. They have been obtained in various ways and analysed. The composition varies with the character of the solution in which the lead compound is formed. Thus in the opalescent or milky solutions in which a proportion of cellulose nitrate is held in solution or semi-solution by the acetone still present, the lead acetate causes a dense coagulation. The precipitates dried and analysed showed 16-20 p.ct. PbO and 11-9 p.ct. N. It is clear that the cellulose nitrates are associated in these precipitates with the lead salts of the acid compounds in question. When the latter are obtained from clear solutions, i.e. in absence of cellulose nitrates, they contain 60-63 p.ct. PbO and 3.5 p.ct. N (obtained as NO).

In further confirmation of the conclusion from these results, viz. that the nitrocelluloses with no tendency to combine with PbO are associated with acid products or by-products of the ester reaction combining with the oxide, the lead reagent was allowed to react in the presence of 90 p.ct. acetone. Water was added, the disintegrated mass collected, washed with dilute acetic acid, and finally with water. Various estimations of the PbO fixed in this way have given numbers varying from 2 to 2.5 p.ct. Such products are perfectly stable. This particular effect of stabilisation appears, therefore, to depend upon the combination of certain acid products present in ordinary nitrocelluloses with metallic oxides. In order to further verify this conclusion, standard specimens of cellulose nitrates have been treated with a large number of metallic salts under varying conditions of action. It has been finally established (1) that the effects in question are more particularly determined by treatment with salts of lead and zinc, and (2) that the simplest method of treatment is that of boiling the cellulose nitrates with dilute aqueous solutions of salts of these metals, preferably the acetates. [Pg 50] The following results may be cited, obtained by boiling a purified 'service' guncotton (sample C) with a 1 p.ct. solution of lead acetate and of zinc acetate respectively. After boiling 60 minutes the nitrates were washed free from the soluble metallic salts, dried and tested.

	Heat Test at 80°	Heat Test at 134°
Original sample C	10	41
Treated with lead acetate	67	45
" zinc "	91	45

In conclusion we may briefly resume the main points arrived at in these investigations.

Causes of instability of cellulose nitrates.—The results of our experiments so far as to the causes of instability in cellulose nitrates may be summed up as follows:—

(1) Traces of free nitrating acids, which can only occur in the finished products through careless manufacture, will undoubtedly cause instability, indicated strongly by the ordinary heat test at 80°, and to a less extent by the heat test at 134°.

(2) Other compounds exist in more intimate association with the cellulose nitrates causing instability which cannot be removed by exhaustive washing with either hot or cold water, by digestion in cold dilute alkaline solutions such as sodium carbonate, or by extracting with ether, alcohol, benzene, &c.; these compounds, however, are soluble in the solvents of highly nitrated cellulose such as acetone, acetic ether, pyridine, &c., even when these liquids are so diluted with water or other non-solvent liquids to such an extent that they have little or no solvent action upon the cellulose nitrate itself. These solutions containing the bodies causing instability are neutral to test paper, but become acid upon evaporation by heating. (This probably explains the presence of free acid when guncotton is [Pg 51] purified by long-continued boiling in water without any neutralising agent being present.)

(3) The bodies causing instability are products or by-products of the original ester reaction, acid bodies containing nitroxy-groups, but otherwise of ill-defined characteristics. They combine with the oxides of zinc or lead, giving insoluble compounds. They are precipitated from their solutions in diluted acetone upon the addition of soluble salts of these metals.

(4) Cellulose nitrates are rendered stable either by eliminating these compounds, or by combining them with the oxides of lead or zinc whilst still in association with cellulose nitrates.

(5) Even the most perfectly purified nitrocellulose will slowly decompose with formation of unstable acid products by boiling for a long time in water. This effect is much more apparent at higher temperatures.

Dense structureless or non-fibrous cellulose nitrates can be industrially prepared (1) by nitrating the amorphous forms of cellulose obtained from its solution as sulphocarbonate (viscose). The cellulose in this condition reacts with the closest similarity to the original fibrous cellulose; the products are similar in composition and properties, including that of instability.

(2) By treating the fibrous cellulose nitrates with liquid solvents of the high nitrate diluted with non-solvent liquids, and more especially water. The optimum effect is a specific disintegration or breaking down of their fibrous structure quite distinct from the gelatinisation which precedes solution in the undiluted solvent, and occurring within narrow limits of variation in the proportion of the diluting and non-solvent liquid—for industrial work the most convenient solution to employ is acetone diluted with about 10 p.ct. of water by volume.

The industrial applications of these results are the basis of English patents 5286 (1898), 18,868 (1898), 18,233 (1898), Luck and Cross (this Journal, 1899, 400, 787). [Pg 52]

The structureless guncotton prepared as above described is of quite exceptional character, and entirely distinct from the ordinary fibrous nitrate or the nitrate prepared by precipitation from actual solution in an undiluted solvent. [3] By the process described, the nitrate is obtained at a low cost in the form of a very fine, dense, structureless, white powder of great purity and stability, entirely free from all mechanical impurities. The elimination of these mechanical impurities, and also to a very great extent of coloured compounds contained in the fibrous nitrate, makes the product also useful in the manufacture of celluloids, artificial silk, &c., whilst its very dense form gives it a great advantage over ordinary fibrous guncotton for use in shells and torpedoes, and for the manufacture of gelatinised gunpowders, &c. It can be compressed with ease into hard masses; and experiments are in progress with a view of producing from it, in admixture with 'retaining' ingredients, a military explosive manufactured by means of ordinary black gunpowder machinery and processes.

Manufacture of sporting powder.—The fact that the fibrous structure of ordinary guncotton or other cellulose nitrate can be completely or partially destroyed by treatment with diluted acetone and without attendant solution, constitutes a process of value for the manufacture of sporting powder having a base of cellulose nitrate of any degree of nitration. The following is a description of the hardening process.

'Soft grains' are manufactured from ordinary guncotton or other cellulose nitrate either wholly or in combination with other ingredients, the process employed being the usual one of revolving in a drum in the damp state and sifting out the grains of suitable size after drying. These grains are then treated with diluted acetone, the degree of dilution being [Pg 53] fixed according to the hardness and bulk of the finished grain it is desired to produce (J. Soc. Chem. Ind., 1899, 787). Owing to the wide limits of dilution and corresponding effect, the process allows of the production of either a 'bulk' or a 'condensed' powder.

We prefer to use about five litres of the liquid to each one kilo. of grain operated upon, as this quantity allows of the grains being freely suspended in the liquid upon stirring. The grains are run into the liquid, which is then preferably heated to the boiling-point for a few minutes whilst the whole is gently stirred. Under this treatment the grains assume a more or less rounded gelatinous condition according to the strength of the liquid. There is, however, no solution of the guncotton and practically no tendency of the grains to cohere. Each grain, however, is acted upon *throughout* and perfectly *equally*. After a few minutes' treatment, water is gradually added, when the grains rapidly harden. They are then freed from acetone and certain impurities by washing with water, heating, and drying. The process is of course carried out in a vessel provided with any means for gentle stirring and heating, and with an outlet for carrying off the volatilised solvent which is entirely recovered by condensation, the grains parting with the acetone with ease.

Stabilising cellulose nitrates.—The process is of especial value in rendering stable and inert the traces of unstable compounds which always remain in cellulose nitrate after the ordinary boiling and washing process. It is of greatest value in the manufacture of collodion cotton used for the preparation of gelatinous blasting explosives and all explosives composed of nitroglycerin and cellulose nitrates. Such mixtures seem peculiarly liable to decomposition if the cellulose nitrate is not of exceptional stability (J. Soc. Chem. Ind., 1899, 787). [Pg 54]

EMPLOI DE LA CELLULOSE POUR LA FABRICATION DE FILS BRILLANTS IMITANT LA SOIE.

E. Bronnert (1) (Rev. Mat. Col., 1900, September, 267).

V. USE OF CELLULOSE IN THE MANUFACTURE OF IMITATIONS OF SILK (LUSTRA-CELLULOSE).

(p. 45) *Introduction.* — The problem of spinning a continuous thread of cellulose has received in later years several solutions. Mechanically all resolve themselves into the preparation of a structureless filtered solution of cellulose or a cellulose derivative, and forcing through capillary orifices into some medium which either absorbs or decomposes the solvent. The author notes here that the fineness and to a great extent the softness of the product depends upon the dimensions of the capillary orifice and concentration of the solution. The technical idea involved in the spinning of artificial fibres is an old one. Réaumur (2) forecast its possibility, Audemars of Lausanne took a patent as early as 1855 (3) for transforming nitrocellulose into fine filaments which he called 'artificial silk.' The idea took practical shape only when it came to be used in connection with filaments for incandescent lamps. In this connection we may mention the names of the patentees: — Swinburne (4), Crookes, Weston (5), Swan (6), and Wynne and Powell (7). These inventors prepared the way for Chardonnet's work, which has been followed since 1888 with continually increasing success.

At this date the lustra-celluloses known may be divided into four classes.

1. 'Artificial silks' obtained from the nitrocelluloses.

2. 'Lustra-cellulose' made from the solution of cellulose in cuprammonium.

3. 'Lustra-cellulose' prepared from the solution of cellulose in chloride of zinc. [Pg 55]

4. 'Viscose silks,' by the decomposition of sulphocarbonate of cellulose (Cross and Bevan).

Group 1. The early history of the Chardonnet process is discussed and some incidental causes of the earlier failures are dealt with. The process having been described in detail in so many publications the reader is referred to these for details. [See Bibliography, (1) and (2), (3) and (4).] The denitrating treatment was introduced in the period 1888-90 and of course altogether changed the prospects of the industry; not only does it remove the high inflammability, but adds considerably to softness, lustre, and general textile quality. In Table I will be found some important constants for the nitrocellulose fibre; also the fibre after denitration and the comparative constants for natural silk.

Table 1.

	Tenacity (grammes)	Elasticity (% elongation)
Nitrocellulose according to Chardonnet German Patent No. 81,599	150	23
The same after denitration	110	8
Denitrated fibre moistened with water	25	—
Nitrocellulose: Bronnert's German Patent No. 93,009	125	28
The same after denitration (dry)	115	13
The same after denitration (wetted)	32	—
Natural silk	300	18

1. Tenacity is the weight in grammes required to break the thread.

2. Elasticity is the elongation per cent. at breaking.

The numbers are taken for thread of 100 deniers (450 metres of 0.05 grammes = 1 denier). It must be noted that according to the concentration of the solution and variations in the process of denitration the constants for the yarn are subject to very considerable variation.

In regard to the manufacture a number of very serious difficulties have been surmounted. First, instead of drying the nitrated cellulose, which often led to fires, &c., it was found better to take it moist from the centrifugal machine, in which [Pg 56] condition it is dissolved (5). It was next found that with the concentrated collodion the thread could be spun direct into the air, and the use of water as a precipitant was thus avoided.

With regard to denitration which is both a delicate and disagreeable operation: none of the agents recommended to substitute the sulphydrates have proved available. Of these the author mentions ferrous chloride (6), ferrous chloride in alcohol (7), formaldehyde (8), sulphocarbonates. The different sulphydrates (9) have very different effects. The calcium compound tends to harden and weaken the thread. The ammonia compound requires great care and is costly. The magnesium compound works rapidly and gives the strongest thread. Investigations have established the following point. In practice it is not necessary to combine the saponification of cellulose ester with complete reduction of the nitric acid split off. The latter requires eight molecules of hydrogen sulphide per one molecule tetranitrocellulose, but with precautions four molecules suffice. It is well known that the denitration is nearly complete, traces only of nitric groups surviving. Their reactions with diphenylamine allow a certain identification of artificial silks of this class. Various other inventors, e.g. Du Vivier (10), Cadoret (11), Lehner (12), have attempted the addition of other substances to modify the thread. These have all failed. Lehner, who persisted in his investigations, and with success, only attained this success, however, by leaving out all such extraneous matters. Lehner works with 10 p.ct. solutions; Chardonnet has continually aimed at higher concentration up to 20 p.ct. Lehner has been able very much to reduce his pressures of ejection in consequence; Chardonnet has had to increase up to pressures of 60 k. per cm. and higher. The latter involves very costly distributing apparatus. Lehner made next considerable advance [Pg 57] by the discovery of the fact that the addition of sulphuric acid to the collodion caused increase of fluidity (13), which Lehner attributes to molecular change. Chardonnet found similar results from the addition of aldehyde and other reagents (14), but not such as to be employed for the more concentrated collodions. The author next refers to his discoveries (15) that alcoholic solutions of a number of substances, organic and inorganic, freely dissolve the lower cellulose nitrates. The most satisfactory of these substances is chloride of calcium (16). It is noted that acetate of ammonia causes rapid changes in the solution, which appear to be due to a species of hydrolysis. The result is sufficiently remarkable to call for further investigation. The chloride of calcium, it is thought possible, produces a direct combination of the alcohol with a reactive group of the nitrocellulose. The fluidity of this solution using one mol. $CaCl_2$ per 1 mol. tetranitrate (17) reaches a maximum in half an hour's heating at 60°-70°C. The fluidity is increased by starting from a cotton which has been previously mercerised. After nitration there is no objection to a chlorine bleach. Chardonnet has found on the other hand that in bleaching before nitration there is a loss of spinning quality in the collodion. The author considers that the new collodion can be used entirely in place of the ordinary ether-alcohol collodion. With regard to the properties of the denitrated products they fix all basic colours without mordant and may be regarded as oxycellulose therefore. The density of the thread is from 1.5 to 1.55. The thread of 100 deniers shows a mean breaking strain of 120 grammes with an elasticity of 8-12 p.ct. The cardinal defect of these fibres is their property of combination with water. Many attempts have been made to confer water-resistance (18), but without success. Strehlenert has proposed the addition of formaldehyde (19), but this is without result (20). In reference to these effects of hydration, the author has made observations [Pg 58] on cotton thread, of which the following table represents the numerical results:

	Breaking Strain Mean of 20 experiments
Skein of bleached cotton without treatment	825
Skein of bleached cotton without treatment, but wetted	942
Ditto after conversion into hexanitrate, dry	884
The above, wetted	828
The cotton denitrated from above, dry	529
The cotton denitrated as above and wetted	206

The author considers that other patents which have been taken for spinning nitrocellulose are of little practical account (21) and (22). The same conclusion also applies to the process of *Langhans*, who proposes to spin solutions of cellulose in sulphuric acid (23) (24) and mixtures of sulphuric acid and phosphoric acid.

Group 2. *Lustra-cellulose.* — Thread prepared by spinning solutions of cellulose in cuprammonium.

This product is made by the Vereinigte Glanzstoff-Fabriken, Aachen, according to a series of patents under the names of H. Pauly, M. Fremery and Urban, Consortium mulhousien pour la fabrication de fils brillants, E. Bronnert, and E. Bronnert and Fremery and Urban (1). The first patent in this direction was taken by Despeissis in 1890 (2). It appears this inventor died shortly after taking the patent (3) The matter was later developed by Pauly (4) especially in overcoming the difficulty of preparing a solution of sufficient concentration. (It is to be noted that Pauly's patents rest upon a very slender foundation, being anticipated in every essential detail by the previous patent of Despeissis.) For this very great care is required, especially, first, the condition of low temperature, and, secondly, a regulated proportion of copper and ammonia to cellulose. The solution takes place more rapidly if the cellulose has been previously oxidised. Such cellulose gives an 8 p.ct. solution, and the thread obtained has the character of an oxycellulose, specially seen in its dyeing properties. The best [Pg 59] results are obtained, it appears, by the preliminary mercerising treatment and placing the alkali cellulose in contact with copper and ammonia. (All reagents employed in molecular proportions.) The author notes that the so-called hydrocellulose (Girard) (5) is almost insoluble in cuprammonium, as is starch. It is rendered soluble by alkali treatment.

Group 3. *Lustra-cellulose* prepared by spinning a solution of cellulose in concentrated chloride of zinc.

This solution has been known for a long time and used for making filaments for incandescent lamps. The cellulose threads, however, have very little tenacity. This is no doubt due to the conditions necessary for forming the solution, the prolonged digestion causing powerful hydrolysis (1). Neither the process of Wynne and Powell (2) nor that of Dreaper and Tompkins (3), who have endeavoured to bring the matter to a practical issue, are calculated to produce a thread taking a place as a textile. The author has described in his American patent (4) a method of effecting the solution in the cold, viz. again by first mercerising the cellulose and washing away the caustic soda. This product dissolves in the cold and the solution remains unaltered if kept at low temperature. Experiments are being continued with these modifications of the process, and the author anticipates successful results. The modifications having the effect of maintaining the high molecular weight of the cellulose, it would appear that these investigations confirm the theory of Cross and Bevan that the tenacity of a film or thread of structureless regenerated cellulose is directly proportional to the molecular weight of the cellulose, i.e. to its degree of molecular aggregation (5).

Group 4. 'Viscose' silks obtained by spinning solutions of xanthate of cellulose.

In 1892, Cross and Bevan patented the preparation of a new and curious compound of cellulose, the thiocarbonate (1) (2) (3). Great hopes were based upon this product at the [Pg 60] time of its discovery. It was expected to yield a considerable industrial and financial profit and also to contribute to the scientific study of cellulose. The later patents of C. H. Stearn (4) describe the application of viscose to the spinning of artificial silk. The viscose is projected into solutions of chloride of ammonium and washed in a succession of saline solutions to remove the residual sulphur impurities. The author remarks that though it has a certain interest to have succeeded in making a thread from this compound and thus adding another to the processes existing for this purpose, he is not of opinion that it shows any advance on the lustra-cellulose (2) and (3). He also considers that the bisulphide of carbon, which must be regarded as a noxious compound, is a serious bar to the industrial use of the process, and for economic work he considers that the regeneration of ammonia from the precipitating liquors is necessary and would be as objectionable as the denitration baths in the collodion process. The final product not being on the market he does not pronounce a finally unfavourable opinion.

The author and the Vereinigte Glanzstoff-Fabriken after long investigation have decided to make nothing but the lustra-cellulose (2) and (3). A new factory at Niedermorschweiler, near Mulhouse, is projected for this last production.

BIBLIOGRAPHY

Introduction

(1) Bull. de la Soc. industr. de Mulhouse, 1900.

(2) Réaumur, Mémoire pour servir à l'histoire des insectes, 1874, 1, p. 154.

(3) English Pat. No. 283, Feb. 6, 1855.

(4) Swinburne, Electrician, 18, 28, 1887, p. 256.

(5) Weston (Swinburne), Electrician, 18, 1887, p. 287. Eng. Pat. No. 22866, Sept. 12, 1882.

(6) German Pat. No. 3029. English Pat. No. 161780, April 28, 1884 (Swan).

(7) Wynne-Powell, English Pat. No. 16805, Dec. 22, 1884.

Group I

(1) German Pat No. 38368, Dec. 20, 1885. German Pat. No. 46125, March 4, 1888. German Pat. No. 56331, Feb. 6, 1890. German Pat. No. 81599, Oct. 11, 1893. German Pat. No. 56655, April 23, 1890. French Pat. No. 231230, June 30, 1893. [Pg 61]

(2) Industrie textile, 1899, 1892. Wyss-Noef, Zeitschrift für angewandte Chemie, 1899, 30, 33. La Nature, Jan. 1, 1898, No. 1283. Revue générale des sciences, June 30, 1898.

(3) German Pat. No. 46125, March 4, 1888. German Pat. No. 56655, April 23, 1890.

(4) Swan, English Pat. 161780, June 28, 1884. See also Béchamp, Dict. de Chimie de Wurtz.

(5) German Pat. No. 81599, Oct 11, 1893.

(6) Béchamp, art. Cellulose, Dict. de Chimie de Wurtz, p. 781.

(7) Chardonnet, addit. March 3, 1897, to the French Pat. 231230, May 30, 1893.

(8) Knofler, French Pat. 247855, June 1, 1895. German Pat. 88556, March 28, 1894.

(9) Béchamp, art. Cellulose, Dict. de Chimie de Wurtz. Blondeau, Ann. Chim. et Phys. (3), 1863, 68, p. 462.

(10) Revue industrielle, 1890, p. 194. German Pat. 52977, March 7, 1889.

(11) French Pat. 256854, June 2, 1896.

(12) German Pat. 55949, Nov. 9, 1889. German Pat. 58508, Sept. 16, 1890. German Pat. 82555, Nov. 15, 1894.

(13) German Pat. 58508, Sept. 16, 1900.

(14) French Pat. 231230, June 30, 1893.

(15) German Pat. 93009, Nov. 19, 1895. French Pat. 254703, March 12, 1896. English Pat. 6858, March 28, 1896.

(16) American Pat. 573132, Dec. 15, 1896.

(17) This proportion is the most advantageous, and furnishes the best liquid collodions that can be spun.

(18) French Pat. 259422, Sept. 3, 1896.

(19) English Pat. 22540, 1896.

(20) Application for German Pat. not granted, 4933 IV. 296, Mar. 16, 1897.

(21) German Pat. 96208, Feb. 10, 1897. Addit. Pat. 101844 and 102573, Dec. 10, 1897.

(22) Oberle et Newbold, French Pat. 25828, July 22, 1896. Granquist, Engl. applic. 2379, Nov. 28, 1899.

(23) German Pat. 72572, June 17, 1891.

(24) Voy. Stern, Ber., 28, ch. 462.

Group II

(1) German Pat. 98642, Dec. 1, 1897 (Pauly). French Pat. 286692, March 10, 1899, and addition of October 14, 1899 (Fremery and Urban). French Pat. 286726, March 11, 1899, and addition of December 4, 1899. German Pat. 111313, March 16, 1899 (Fremery and Urban). English Pat. 18884, Sept. 19, 1899 (Bronnert). English Pat. 13331, June 27, 1899 (Consort. mulhousien).

(2) French Pat. 203741, Feb. 12, 1890.

(3) The actual lapse of this patent is due to the death of Despeissis shortly after it was taken. [Pg 62]

(4) Without questioning the good faith of Pauly, it is nevertheless a fact that the original patent remains as a document, and therefore that the value of the Pauly patents is very questionable.

(5) Girard, Ann. Chim. et Phys, 1881 (5), 24, p. 337-384.

Group III

(1) Cross and Bevan, Cellulose, 1895, p. 8.

(2) English Pat. 16805, Dec. 22, 1884.

(3) English Pat. 17901, July 30, 1897.

(4) Bronnert, American Pat. 646799, April 3, 1900.

(5) Cross and Bevan, Cellulose, 1895, p. 12.

Group IV

(1) English Pat. 8700, 1892. German Pat. 70999, Jan. 13, 1893.

(2) English Pat. 4713, 1896. German Pat. 92590, Nov. 21, 1896.

(3) Comptes rendus (loc. cit.). Berichte, c. 9, 65a.

(4) English Pat. 1020, 1898. German Pat. 108511, Oct. 18, 1898.

Artificial Silk—Lustra-cellulose.

C. F. Cross and E. J. Bevan (J. Soc. Chem. Ind., 1896, 317).

The object of this paper is mainly to correct current statements as to the artificial or 'cellulose silks' being explosive or highly inflammable (*ibid.*, 1895, 720). A specimen of the 'Lehner' silk was found to retain only 0.19 p.ct. total nitrogen, showing that the denitration is sufficiently complete to dispose of any suggestion of high inflammability.

The product yielded traces only of furfural; on boiling with a 1 p.ct. solution of sodium hydrate, the loss of weight was 9.14 p.ct.; but the solution had no reducing action on Fehling's solution. The product in denitration had therefore reverted completely to a cellulose (hydrate), no oxy-derivative being present.

The authors enter a protest against the term 'artificial silk' as applied to these products, and suggest 'lustra-cellulose.' [Pg 63]

DIE KÜNSTLICHE SEIDE-IHRE HERSTELLUNG, EIGENSCHAFTEN UND VERWENDUNG.

Carl Süvern, Berlin, 1900, J. Springer.

ARTIFICIAL SILK—ITS PRODUCTION, PROPERTIES, AND APPLICATIONS.

This work of some 130 pages is an important monograph on the subject of the preparation of artificial cellulose threads—so far as the technical elements of the problems involved are discussed and disclosed in the patent literature. The first section, in fact, consists almost exclusively of the several patent specifications in chronological order and ranged under the sub-sections: (*a*) The Spinning of Nitrocellulose (collodion); (*b*) The Spinning of other Solutions of Cellulose; (*c*) The Spinning of Solutions of the Nitrogenous Colloids.

In the second section the author deals with the physical and chemical proportions of the artificial threads.

Chardonnet '*silk*' is stated to have a mean diameter of 35μ, but with considerable variations from the mean in the individual fibres; equally wide variations in form are observed in cross-section. The general form is elliptical, but the surface is marked by deep striæ, and the cross-section is therefore of irregular outline. This is due to irregular conditions of evaporation of the solvents, the thread being 'spun' into the air from cylindrical orifices of regulated dimensions. Chardonnet states that when the collodion is spun into alcohol the resultant thread is a perfect cylinder (Compt. rend. 1889, 108, 962). The strength of the fibre is variously stated at from 50-80 p.ct. that of 'boiled off' China tram; the true elasticity is 4-5 p.ct., the elongation under the breaking strain 15-17 p.ct. The sp.gr. is 1.49, i.e. 3-5 p.ct. in excess of boiled off silk.

Lehner 'silk' exhibits the closest similarity to the Chardonnet [Pg 64] product. In cross-section it is seen to be more regular in outline, and a round, pseudo-tubular form prevails, due to the conditions of shrinkage and collapse of the fibre in parting with the solvents, and in then dehydrating. The constants for 'breaking strain,' both in the original and moistened condition, for elasticity, &c., are closely approximate to those for the Chardonnet product.

Pauly 'silk'. — The form of the ultimate fibres is much more regular and the contour of the cross-section is smooth. The product shows more resistance to moisture and to alkaline solutions.

Viscose 'silk' is referred to in terms of a communication appearing in 'Papier-Zeitung,' 1898, 2416.

In the above section the following publications are referred to: Chardonnet, 'Compt. rend.,' 1887, 105, 900; and 1889, 108, 962; Silbermann, 'Die Seide,' 1897, v. 2, 143; Herzog, 'Farber-Zeitung,' 1894/5, 49-50; Thiele, ibid. 1897, 133; O. Schlesinger, 'Papier-Zeitung,' 1895, 1578-81, 1610-12.

Action of Reagents upon Natural and Artificial Silks.

1. *Potassium hydrate* in solution of maximum concentration dissolves the silks proper, (*a*) China silk on slight warming, (*b*) Tussah silk on boiling. The cellulose 'silks' show swelling with discolouration, but the fibrous character is not destroyed even on boiling.

2. *Potassium hydrate* 40 p.ct. China silk dissolves completely at 65°-85°; Tussah silk swells considerably at 75° and dissolves at 100°-120°. The cellulose 'silks' are attacked with discolouration; at 140° (boiling-point of the solution) there is progressive solvent action, but the action is incomplete. The Pauly product is most resistant.

3. *Zinc chloride*, 40 p.ct. solution. Both the natural silks and lustra-celluloses are attacked at 100°, and on raising the temperature the further actions are as follows: China silk is [Pg 65] completely dissolved at 110-120°; Tussah silk at 130-135°; the collodion products at 140-145°; the Pauly product was again most resistant, dissolving at 180°.

4. *Alkaline cupric oxide* (glycerin) solution was prepared by dissolving 10 grs. of the sulphate in 100 c.c. water, adding 5 grs. glycerin and 10 c.c. of 40 p.ct. KOH. In this solution the China silk dissolved at the ordinary temperature; Tussah silk and the lustra-celluloses were not appreciably affected.

5. *Cuprammonium solution* was prepared by dissolving the precipitated cupric hydrate in 24 p.ct. ammonia. In this reagent also the China silk dissolved, and the Tussah silk as well as the lustra-celluloses underwent no appreciable change.

6. *An ammoniacal solution of nickel oxide* was prepared by dissolving the precipitated hydrated oxide in concentrated ammonia. The China silk was dissolved by this reagent; Tussah silk and the lustra-celluloses entirely resisted its action.

7. *Fehling's solution* is a solvent of the natural silks, but is without action on the lustra-celluloses.

8. *Chromic acid* — 20 p.ct. CrO_3 — solution dissolves both the natural silks and the lustra-celluloses at the boiling temperature of the solution.

9. *Millon's reagent*, at the boiling solution, colours the natural silks violet: the lustra-celluloses give no reaction.

10. *Concentrated nitric acid* attacks the natural silks powerfully in the cold; the lustra-celluloses dissolve on heating.

11. *Iodine solution* (I in KI) colours the China silk a deep brown, Tussah a pale brown; the celluloses from collodion are coloured at first brown, then blue. The Pauly product, on the other hand, does not react.

12. *Diphenylamine sulphate.* — A solution of the base in concentrated sulphuric acid colours the natural silks a brown; the collodion 'silks' give a strong blue reaction due to the [Pg 66] presence of residual nitro-groups. The Pauly product is not affected.

13. *Brucin sulphate* in presence of concentrated sulphuric acid colours the natural silks only slightly (brown); the collodion 'silks' give a strong red colouration. The Pauly product again is without reaction.

14. *Water.* — The natural silks do not soften in the mouth as do the lustra-celluloses.

15. *Water of condition* was determined by drying at 100°; the following percentages resulted (*a*). The percentages of water (*b*) taken up from the atmosphere after forty-three hours' exposure were:

	(*a*)	(*b*)
China (raw) silk	7.97	2.24
Tussah silk	8.26	5.00
Lustra-celluloses:		
Chardonnet (Besançon)	10.37	5.64
" Spreitenbach	11.17	5.77
Lehner	10.71	5.97
Pauly	10.04	6.94

16. *Behaviour on heating at 200°.* — After two hours' heating at this temperature the following changes were noted:

China silk	Much discoloured (brown).
Tussah silk	Scarcely affected.
Lustra-celluloses:	
Chardonnet Lehner	Converted into a blue-black charcoal, retaining the form of the fibres.
Pauly	A bright yellow-brown colouration, without carbonisation.

17. The *losses of weight* accompanying these changes and calculated per 100 parts of fibre dried at 100° were:

China silk	3.18
Tussah silk	2.95

[Pg 67]

Lustra-celluloses:	
Chardonnet	33.70
Lehner	26.56
Pauly	1.61

18. *Inorganic constituents.*—Determinations of the total ash gave for the first five of the above, numbers varying from 1.0 to 1.7 p.ct. The only noteworthy point in the comparison was the exceptionally small ash of the Pauly product, viz. 0.096 p.ct.

19. *Total nitrogen.*—The natural silks contain the 16-17 p.ct. N characteristic of the proteids. The lustra-celluloses contain 0.05-0.15 p.ct. N which in those spun from collodion is present in the form of nitric groups.

The points of chemical differentiation which are established by the above scheme of comparative investigation are summed up in tabular form.

Methods of dyeing.—The lustra-celluloses are briefly discussed. The specific relationship of these forms of cellulose to the colouring matters are in the main those of cotton, but they manifest in the dye-bath the somewhat intensified attraction which characterises mercerised cotton, or more generally the cellulose hydrates.

Industrial applications of the lustra-celluloses are briefly noticed in the concluding section of the book.

FOOTNOTES:

[3] With these products it is easy to observe that they have a definite fusion point 5°-10° below the temperature of explosion.

SECTION III. DECOMPOSITIONS OF CELLULOSE SUCH AS THROW LIGHT ON THE PROBLEM OF ITS CONSTITUTION

UEBER CELLULOSE.

G. Bumcke und R. Wolffenstein (Berl. Ber., 1899, 2493).

(p. 54) *Theoretical Preface.* — The purpose of these investigations is the closer characterisation of the products known as 'oxycellulose' and 'hydracellulose,' which are empirical aggregates obtained by various processes of oxidation and [Pg 68] hydrolysis; these processes act concurrently in the production of the oxycelluloses. The action of hydrogen peroxide was specially investigated. An oxycellulose resulted possessing strongly marked aldehydic characteristics. The authors commit themselves to an explanation of this paradoxical result, *i.e.* the production of a body of strongly 'reducing' properties by the action of an oxidising agent upon the inert cellulose molecule (? aggregate) as due to the *hydrolytic* action of the peroxide: following Wurster (Ber. 22, 145), who similarly explained the production of reducing sugars from cane sugar by the action of the peroxide.

The product in question is accordingly termed *hydralcellulose*. By the action of alkalis this is resolved into two bodies of alcoholic (cellulose) and acid ('acid cellulose') characteristics respectively. The latter in drying passes into a lactone. The acid product is also obtained from cellulose by the action of alkaline lye (boiling 30 p.ct. NaOH) and by solution in Schweizer's reagent.

It is considered probable that the cellulose nitrates are hydrocellulose derivatives, and experimental evidence in favour of this conclusion is supplied by the results of 'nitrating' the celluloses and their oxy- and hydro- derivatives. Identical products were obtained.

Experimental investigations. — The filter paper employed as 'original cellulose,' giving the following numbers on analysis:

C	44.56	44.29	44.53	44.56
H	6.39	6.31	6.46	6.42

was exposed to the action of pure distilled H_2O_2 at 4-60 p.ct. strength, at ordinary temperatures until disintegrated: a result requiring from nineteen to thirty days. The series of products gave the following analytical results:

C	43.61	43.61	43.46	43.89	44.0	43.87	43.92	43.81
H	6.00	6.29	6.28	6.26	6.13	6.27	6.24	6.27

[Pg 69]

results lying between the requirements of the formulæ:

$5\ C_6H_{10}O_5.H_2O$ and $8\ C_6H_{10}O_5.H_2O$.

Hydrazones were obtained with 1.7-1.8 p.ct. N. Treated with caustic soda solution the hydrazones were dissolved in part: on reprecipitation a hydrazone of unaltered composition was obtained. The original product shows therefore a uniform distribution of the reactive CO- groups.

The hydralcellulose boiled with Fehling's solution reduced 1/12 of the amount required for an equal weight of glucose.

Digested with caustic soda solution it yielded 33 p.ct. of its weight of the soluble 'acid cellulose.' This product was purified and analysed with the following result: C 43.35 H 6.5. For the direct production of the 'acid' derivative, cellulose was boiled with successive quantities of 30 p.ct. NaOH until *dissolved*. It required eight treatments of one hour's duration. On adding sulphuric acid to the solutions the product was precipitated. Yield 40 p.ct. Analyses:

C	43.8	43.8	43.7
H	6.2	6.2	6.3

The cellulose reprecipitated from solution in Schweizer's reagent gave similar analytical results:

C	43.9	43.8	44.0

| H | 6.5 | 6.3 | 6.4 |

Conversion into nitrates. — The original cellulose, hydral- and acid cellulose were each treated with 10 times their weight of HNO_3 of 1.48 sp.gr. and heated at 85° until the solution lost its initial viscosity.

The products were precipitated by water and purified by solution in acetone from which two fractions were recovered, the one being relatively insoluble in ethyl alcohol. The [Pg 70] various nitrates from the several original products proved to be of almost identical composition,

C 32.0   H 4.2   N 8.8

with a molecular weight approximately 1350. The conclusion is that these products are all derivatives of a 'hydralcellulose' 6 $C_6H_{10}O_5H_2O$.

FORMATION OF FURFURALDEHYDE FROM CELLULOSE, OXYCELLULOSE, AND HYDROCELLULOSE.

By Leo Vignon (Compt. rend., 1898, 126, 1355-1358).

(p. 54) Hydrocellulose, oxycellulose, and 'reduced' cellulose, the last named being apparently identical with hydrocellulose, were obtained by heating carefully purified cotton wool (10 grams) in water (1,000 c.c.), with (1) 65 c.c. of hydrochloric acid (1.2 sp.gr.), (2) 65 c.c. of hydrochloric acid and 80 grams of potassium chlorate, (3) 65 c.c. of hydrochloric acid and 50 grams of stannous chloride. From these and some other substances, the following percentage yields of furfuraldehyde were obtained: Hydrocellulose, 0.854; oxycellulose, 2.113; reduced cellulose, 0.860; starch, 0.800; bleached cotton, 1.800; oxycellulose, prepared by means of chromic acid, 3.500. Two specimens of oxycellulose were prepared by treating cotton wool with hydrochloric acid and potassium chlorate (A), and with sulphuric acid and potassium dichromate (B), and 25 grams of each product digested with aqueous potash. Of the product A, 16.20 grams were insoluble in potash, 2.45 grams were precipitated on neutralisation of the alkaline solution, and 6.35 grams remained in solution, whilst B yielded 11.16 grams of insoluble matter, 1.42 grams were precipitated by acid, and 12.42 grams remained in solution. The percentage yields of [Pg 71] furfuraldehyde obtained from these fractions were as follows: A, insoluble, 0.86; precipitated, 4.35; dissolved, 1.10. B, insoluble, 0.76; precipitated, 5.11; dissolved, 1.54. It appears, from the foregoing results, that the cellulose molecule, after oxidation, is easily decomposed by potash, the insoluble and larger portion having all the characters of the original cellulose, whilst the soluble portion is of an aldehydic nature, and contains a substance, precipitable by acids, which yields a relatively large amount of furfuraldehyde.

UNTERSUCHUNGEN ÜBER DIE OXYCELLULOSE.

O. v. Faber und B. Tollens (Berl. Ber., 1899, 2589).

Investigations of Oxycellulose.

(p. 61) The author's results are tersely summed up in the following conclusions set forth at the end of the paper: The oxycelluloses are mixtures of cellulose and a derivative oxidised compound which contains one more atom O than cellulose (cellulose = $C_6H_{10}O_5$), and for which the special designation *Celloxin* is proposed.

Celloxin may be formulated $C_8H_6O_6$ or $C_6H_{10}O_6$, of which the former is the more probable.

The various oxycelluloses may be regarded as containing one celloxin group to 1-4 cellulose groups, according to the nature of the original cellulose, and the degree of oxidation to which subjected. These groups are in chemical union.

Celloxin has not been isolated. On boiling the oxycelluloses with lime-milk it is converted into isosaccharinic and dioxybutyric acids. The insoluble residue from the treatment is cellulose. [Pg 72]

The following oxycelluloses were investigated:

A. *Product of action of nitric acid upon pine wood* (Lindsey and Tollens, Ann. 267, 366).—The oxycelluloses contained

1 mol celloxin: {2 mol. cellulose on 6 hours' heating
{3 mol. cellulose on 3 hours' heating

with a ratio H: O = 1: 9 and 1: 8.7 respectively: they yielded 7 p.ct. furfural.

B. *By action of bromine in presence of water and* $CaCO_3$ *upon cotton.*—Yield, (air-dry) 85 p.ct. Empirical composition $C_{12}H_{20}O_{11}$ = $C_6H_{10}O_5.C_6H_{10}O_6$: yielded furfural 1.7 p.ct.

C. *Cotton and nitric acid at* 100°, two and a half hours (Cross and Bevan).—Yield, 70 p.ct. Composition

4 $C_6H_{10}O_5.C_6H_8O_6$

yielded furfural 2.3 p.ct.

D. *Cotton and nitric acid at* 100° (four hours).—A more highly oxidised product resulted, viz. 3 $C_6H_{10}O_5.C_6H_8O_6$: yielded furfural 3.2 p.ct.

By-products of oxidation.—The liquors from B were found to contain saccharic acid: the acid from C and B contained a dibasic acid which appeared to be tartaric acid.

The isolation of (1) isosaccharinic and (2) dioxybutyric acid from the products of digestion of the oxycelluloses with lime-milk at 100° was effected by the separation of their respective calcium salts, (1) by direct crystallisation, (2) by precipitation alcohol after separation of the former.

CELLULOSES, HYDRO- AND OXYCELLULOSES, AND CELLULOSE ESTERS.

L. Vignon (Bull. Soc. Chim., 1901 [3], 25, 130).

(a) *Oxycelluloses from cotton, hemp, flax, and ramie.* — The comparative oxidation of these celluloses, by treatment with $HClO_3$ at 100°, gave remarkably uniform results, as shown by the following numbers, showing extreme variations: yields, 68-70 p.ct.; hydrazine reaction, N fixed 1.58-1.69; fixation of basic colouring matters (relative numbers), saffranine, 100-200, methylene blue, 100-106. The only points of difference noted were (1) hemp is somewhat more resistant to the acid oxidation; (2) the cotton oxycellulose shows a somewhat higher (25 p.ct.) cupric reduction.

(b) *'Saccharification' of cellulose, cellulose hydrates, and hydrocellulose.* — The products were digested with dilute hydrochloric acid six hours at 100°, and the cupric reduction of the soluble products determined and calculated to dextrose.

100 grms. of	gave reducing products equal to Dextrose
Purified cotton	3.29
" Hydrocellulose	9.70
Cotton mercerised (NaOH 30° B.)	4.39
Cotton mercerised (NaOH 40° B.)	3.51
Cellulose reprecipitated from cuprammonium	4.39
Oxycellulose	14.70
Starch	98.6

These numbers show that cellulose may be hydrated both by mercerisation and solution, without affecting the constitutional relationships of the CO groups. The results also differentiate the cellulose series from starch in regard to hydrolysis.

(c) *Cellulose and oxycellulose nitrates.* — The nitric esters of cellulose have a strong reducing action on alkaline copper solutions. The author has studied this reaction quantitatively for the esters both of cellulose and oxycellulose, at two stages of 'nitration,' represented by 8.2-8.6 p.ct. and 13.5-13.9 p.ct. total nitrogen in the ester-products, respectively. The results are expressed in terms (c.c.) of the cupric reagent (Pasteur) reduced per 100 grs. compared with dextrose (=17767). [Pg 74]

Cellulose maximum nitration (13.5 p.ct. N)	3640
Oxycellulose maximum nitration (13.9 p.ct. N)	3600
Cellulose minimum nitration (8.19 p.ct. N)	3700
Oxycellulose minimum nitration (8.56 p.ct. N)	3620

The author concludes that, since the reducing action is independent of the degree of nitration, and is the same for cellulose and the oxycelluloses, the ester reaction in the case of the normal cellulose is accompanied by oxidation, the product being an oxycellulose ester.

Products of 'denitration'.—The esters were treated with ferrous chloride in boiling aqueous solution. The products were oxycelluloses, with a cupric reduction equal to that of an oxycellulose directly prepared by the action of $HClO_3$. On the other hand, by treatment with ammonium sulphide at 35°-40° 'denitrated' products were obtained without action on alkaline copper solutions.

OXYCELLULOSES AND THE MOLECULAR WEIGHT OF CELLULOSE.

H. Nastukoff (Berl. Ber. 33 [13] 2237).

(p. 61) The author continues his investigations of the oxidation of cellulose. [Compare Bull. Mulhouse, 1892.] The products described were obtained by the action of hypochlorites and permanganates upon Swedish filter paper (Schleicher and Schüll).

4. *Oxidation by hypochlorites.*—(1) The cellulose was digested 24 hrs. with 35 times its weight of a filtered solution of bleaching power of 4°B.; afterwards drained and exposed for 24 hrs. to the atmosphere. These treatments were then repeated. After washing, treatment with dilute acetic acid and again washing, the product was treated with a 10 p.ct. NaOH [Pg 75] solution. The oxycellulose was precipitated from the filtered solution: yield 45 p.ct. The residue

when purified amounted to 30 p.ct. of the original cellulose, with which it was identical in all essential properties.

The oxycellulose, after purification, dried at 110°, gave the following analytical numbers:

C	43.64	43.78	43.32	43.13
H	6.17	6.21	5.98	6.08

Its compound with phenylhydrazine (*loc. cit.*) gave the following analytical numbers:

N	0.78	0.96	0.84

(2) The reagents were as in (1), but the conditions varied by passing a stream of carbonic acid gas through the solution contained in a flask, until Cl compounds ceased to be given off. The analysis of the purified oxycellulose gave C 43.53, H 6.13.

(3) The conditions were as in (2), but a much stronger hypochlorite solution—viz. 12°B.—was employed. The yield of oxycellulose precipitated from solution in soda lye (10 p.ct. NaOH) was 45 p.ct. There was only a slight residue of unattacked cellulose. The analytical numbers obtained were:

Oxycellulose	C	43.31	43.74	43.69
"	H	6.47	6.42	6.51
Phenylhydrazine compound	N		0.62	0.81

B. *Oxidation by permanganate* ($KMnO_4$). (1) The cellulose 16 grms. was treated with 1100 c.c. of a 1 p.ct. solution of $KMnO_4$ in successive portions. The MnO_2 was removed from time to time by digesting the product with a dilute sulphuric acid (10 p.ct. H_2SO_4). The oxycellulose was purified as before, yield 40 p.ct. Analytical numbers:

Oxycellulose	C	42.12	42.9
"	H	6.20	6.11

| Phenylhydrazine compound | N | 1.35 | 1.08 | 1.21 |

[Pg 76]

(2) The cellulose (16 grms.) was digested 14 days with 2500 c.c. of 1 p.ct. $KMnO_4$ solution. The purified oxycellulose was identical in all respects with the above: yield 40 p.ct. C 42.66, H 6.19.

(3) The cellulose (16 grms.) was heated in the water-bath with 1600 c.c. of 15 p.ct. H_2SO_4 to which were added 18 grms. $KMnO_4$. The yield and composition of the oxycellulose was identical with the above. It appears from these results that the oxidation with hypochlorites acids 1 atom of O to 4-6 of the unit groups $C_6H_{10}O_5$; and the oxidation with permanganate 2 atoms O per 4-6 units of $C_6H_{10}O_5$. The molecular proportion of N in the phenylhydrazine residue combining is fractional, representing 1 atom O, *i.e.* 1 CO group reacting per 4 $C_{36}H_{60}O_{31}$ and 6 $C_{24}H_{49}O_{21}$ respectively, assuming the reaction to be a hydrazone reaction.

Further investigations of the oxycelluloses by treatment with (*a*) sodium amalgam, (*b*) bromine (water), and (*c*) dilute nitric acid at 110°, led to no positive results.

By treatment with alcoholic soda (NaOH) the products were resolved into a soluble and insoluble portion, the properties of the latter being those of a cellulose (hydrate).

Molecular weight of cellulose and oxycellulose. — The author endeavours to arrive at numbers expressing these relations by converting the substances into acetates by Schutzenberger's method, and observing the boiling-points of their solution in nitrobenzene.

FERMENTATION OF CELLULOSE

V. Omelianski (Compt. Rend., 1897, 125, 1131-1133).

Pure paper was allowed to ferment in the presence of calcium carbonate at a temperature of 35° for 13 months. The [Pg 77] products obtained from 3.4743 grams of paper were: acids of the acetic series, 2.2402 grams; carbonic anhydride, 0.9722 grams; and hydrogen, 0.0138 gram. The acids were chiefly acetic and butyric acid, the ratio of the former to the latter being 1.7: 1. Small quantities of valeric acid, higher alcohols, and odorous products were formed.

The absence of methane from the products of fermentation is remarkable, but the formation of this gas seems to be due to a special organism readily distinguishable from the ferment that produces the fatty acids. This organism is at present under investigation.

(p. 75) **Constitution of Cellulose.** — It may be fairly premised that the problem of the constitution of cellulose cannot be solved independently of that of molecular aggregation. We find in effect that the structural properties of cellulose and its derivatives are directly connected with their constitution. So far we have only a superficial perception of this correlation. We know that a fibrous cellulose treated with acids or alkalis in such a way that only hydrolytic changes can take place is converted into a variety of forms of very different structural characteristics, and these products, while still preserving the main chemical characteristics of the original, show when converted into derivatives by simple synthesis, *e.g.* esters and sulphocarbonates, a corresponding differentiation of the physical properties of these derivatives, from the normal standard, and therefore that the new reacting unit determines a new physical aggregate. Thus the sulphocarbonate of a 'hydrocellulose' is formed with lower proportions of alkaline hydrate and carbon disulphide, gives solutions of relatively low viscosity, and, when decomposed to give a film or thread of the regenerated cellulose, these are found to be deficient in strength [Pg 78] and elasticity. Similarly with the acetate. The normal acetate gives solutions of high viscosity, films of considerable tenacity, and when those are saponified the cellulose is regenerated as an unbroken film. The acetates of hydrolysed celluloses manifest a retrogradation in structural and physical properties, proportioned to the degree of hydrolysis of the original.

We may take this opportunity of pointing out that the celluloses not only suggest with some definiteness the connection of the structural properties of visible aggregates — that is, of matter in the mass — with the configuration of the chemical molecule or reacting unit, but supply unique material for the actual experimental investigation of the problems involved. Of all the 'organic' colloids cellulose is the only one which can be converted into a variety of derivative forms, from each of which a regular solid can be produced in continuous length and of any prescribed dimensions. Thus we can compare the structural properties of cellulose with those of its hydrates, nitrates, acetates, and benzoates, in terms of measurements of breaking strain, extensibility, elasticity. Investigations in this field are being prosecuted, but the results are not as yet sufficiently elaborated for reduction to formulæ. One striking general conclusion is, however, established, and that is that the structural properties of cellulose are but little affected by esterification and appear therefore to be a function of the special arrangement of the carbon atoms, i.e. of the molecular constitution. Also it is established that the molecular aggregate which constitutes a cellulose is of a resistant type, and undoubtedly persists in the solutions of the compounds.

It may be urged that it is superfluous to import these questions of mass-aggregation into the problem of the chemical constitution of cellulose. But we shall find that the point again arises in attempting to define the reacting unit, which is another term for the molecule. In the majority of cases we [Pg 79] rely for this upon physical measurements; and in fact the purely chemical determination of such quantities is inferential. Attempts have been made to determine the molecular weights of the cellulose esters in solution, by observations of depression of solidifying and boiling-points. But the numbers have little value. The only other well-defined compound is the sulphocarbonate. It has been pointed out that, by successive precipitations of this compound, there occurs a continual aggregation of the cellulose with dissociation of the alkali and CS residues and it has been found impossible to assign a limit to the dissociation, i.e. to fix a point at which the transition from soluble sulphocarbonate to insoluble cellulose takes place.

On these grounds it will be seen we are reduced to a somewhat speculative treatment of the hypothetical ultimate unit group, which is taken as of C_6 dimensions.

As there has been no addition of experimental facts directly contributing to the solution of the problem, the material available for a discussion of the probabilities remains very much as stated in the first edition, pp. 75-77. It is now generally admitted that the tetracetate n [$C_6H_6O.(OAc)_4$] is a normal cellulose ester; therefore that four of the five O atoms are hydroxylic. The fifth is undoubtedly carbonyl oxygen. The reactions of cellulose certainly indicate that the CO- group is ketonic rather than aldehydic. Even when attacked by strong sulphuric acid the resolution proceeds some considerable way before products are obtained reducing Fehling's solution. This is not easily reconcilable with any polyaldose formula. Nor is the resistance of cellulose to very severe alkaline treatments. The probability may be noted here that under the action of the alkaline hydrates there occurs a change of configuration. Lobry de Bruyn's researches on the change of position of the typical CO- group of the simple hexoses, in presence of alkalis, point very definitely in this direction. It [Pg 80] is probable that in the formation of alkali cellulose there is a constitutional change of the cellulose, which may in effect be due to a migration of a CO- position within the unit group. Again also we have the interesting fact that structural changes accompany the chemical reaction. It is surprising that there should have been no investigation of these changes of external form and structure, otherwise than as mass effects. We cannot, therefore, say what may be the molecular interpretation of these effects. It has not yet been determined whether there are any intrinsic volume changes in the cellulose substance itself: and as regards what changes are determined in the reacting unit or molecule, we can only note a fruitful subject for future investigation. *A priori* our views of the probable changes depend upon the assumed constitution of the unit group. If of the ordinary carbohydrate type, formulated with an open chain, there is little to surmise beyond the change of position of a CO- group. But alternative formulæ have been proposed. Thus the tetracetate is a derivative to be reckoned with in the problem. It is formed under conditions which preclude constitutional changes within the unit groups. The temperature of the main reaction is 30°-40°, the reagents are used but little in excess of the quantitative proportions, and the yields are approximately quantitative. If now the derivative is formed entirely without the hydrolysis the empirical formula $C_6H_6O.(OAc)_4$ justifies a closed-ring formula for the original viz. $CO<[CHOH]_4>CH_2$; and the preference for this formula depends upon the explanation it affords of the aggregation of the groups by way of CO-CH$_2$ synthesis.

The exact relationship of the tetracetate to the original cellulose is somewhat difficult to determine. The starting-point is a cellulose hydrate, since it is the product obtained by decomposition of the sulphocarbonate. The degree of *hydrolysis* attending the cycle of [Pg 81] reactions is indicated by the formula 4 C$_6$H$_{10}$O$_5$.H$_2$O. It has been already shown that this degree of hydrolysis does not produce molecular disaggregation. If this hydrate survived the acetylation it would of course affect the empirical composition, i.e. chiefly the carbon percentage, of the product. It may be here pointed out that the extreme variation of the carbon in this group of carbohydrate esters is as between C$_{14}$H$_{20}$O$_{10}$ (C = 48.3 p.ct.) and C$_{14}$H$_{18}$O$_9$ (C = 50.8 p.ct.) i.e. a tetracetate of C$_6$H$_{12}$O$_6$ and C$_6$H$_{10}$O$_5$ respectively. In the fractional intermediate terms it is clear that we come within the range of ordinary experimental errors, and to solve this critical point by way of ultimate analysis must involve an extended series of analyses with precautions for specially minimising and quantifying the error. The determination of the acetyl by saponification is also subject to an error sufficiently large to preclude the results being applied to solve the point. While, therefore, we must defer the final statement as to whether the tetracetate is produced from or contains a partly hydrolysed cellulose molecule, it is clear that at least a large proportion of the unit groups must be acetylated in the proportion C$_6$H$_6$O.(OAc)$_4$.

It has been shown that by the method of Franchimont a higher proportion of acetyl groups can be introduced; but this result involves a destructive hydrolysis of the cellulose: the acetates are not derivatives of cellulose, but of products of hydrolytic decomposition.

It appears, therefore, that with the normal limit of acetylation at the tetracetate the aggregation of the unit groups must depend upon the CO- groups and a ring formula of the general form CO<[CHOH]$_4$>CH$_2$ is consistent with the facts.

Vignon has proposed for cellulose the constitutional formula

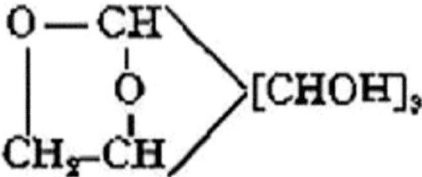

with reference to the highest nitrate, and the decomposition of the nitrate by alkalis with formation of hydroxypyruvic acid. While these reactions afford no very sure ground for deductions as to constitutional [Pg 82] relationships, it certainly appears that, if the aldose view of the unit group is to be retained, this form of the anhydride contains suggestions of the general tendency of the celluloses on treatment with condensing acids to split off formic acid in relatively large quantity [Ber. 1895, 1940]; the condensation of the oxycelluloses to furfural; the non-formation of the normal hydroxydicarboxylic acids by nitric acid oxidations. Indirectly we may point out that any hypothesis which retains the polyaldose view of cellulose, and so fails to differentiate its constitution from that of starch, has little promise of progress. The above formula, moreover, concerns the assumed unit group, with no suggestion as to the mode of aggregation in the cellulose complex. Also there is no suggestion as to how far the formula is applicable to the celluloses considered as a group. In extending this view to the oxycelluloses, Vignon introduces the derived oxidised group

$$CHO.(CHOH)_3.CH \cdot CO\!\!-\!\!o$$
$$\underline{O}$$

—of which one is apportioned to three or four groups of the cellulose previously formulated: these groups in condensed union together constitute an oxycellulose.

These views are in agreement with the experimental results obtained by Faber and Tollens (p. 71). They regard the oxycelluloses as compounds of 'celloxin' $C_6H\_8\{O\}_6$ with 1-4 mols. unaltered cellulose; and the former they particularly refer to as a lactone of glycuronic acid. But on boiling with lime they obtain dioxybutyric and isosaccharinic acids; both of which are not very obviously related to the compounds formulated by Vignon. We revert with preference to a definitely ketonic formula, for which, moreover, some farther grounds remain to be mentioned. In the systematic investigation of the nitric esters [Pg 83] of the carbohydrates (p. 41) Will and Lenze have definitely differentiated the ketoses from the aldoses, as showing an internal condensation accompanying the ester reaction. Not only are the OH groups taking part in the latter consequently less by two than in the corresponding aldoses, but the nitrates show a much increased stability. This would give a simple explanation of the well-known facts obtaining in the corresponding esters of the normal cellulose. We may note here that an important item in the quantitative factors of the cellulose nitric ester reaction has been overlooked: that is, the yield calculated to the NO_3 groups fixed. The theoretical yields for the higher nitrates are

	Yield p.ct. of cellulose	N p.ct. of nitrate
Pentanitrate	169	12.7
Hexanitrate	183	14.1

From such statistics as are recorded the yields are not in accordance with the above. There is a sensible deficiency. Thus Will and Lenze record a yield of 170 p.ct. for a product with 13.8 p.ct. N, indicating a deficiency of about 10 p.ct. As the by-products soluble in the acid mixture are extremely small, the deficiency represents approximately the water split off by an internal reaction. In this important point the celluloses behave as ketoses.

In the lignocelluloses the condensed constituents of the complex are of well-marked ketonic, i.e. quinonic, type. In 'nitrating' the lignocelluloses this phenomenon of internal condensation is much more pronounced (see p. 131). As the reaction is mainly confined to the cellulose of the fibre, we have this additional evidence that the typical carbonyl is of ketonic function. It is still an open question whether the cellulose constituents of the lignocelluloses are progressively condensed — with progress of 'lignification' — to the unsaturated [Pg 84] or lignone groups. There is much in favour of this view, the evidence being dealt with in the first edition, p. 180. The transition from a cellulose-ketone to the lignone-ketone involves a simple condensation without rearrangement; from which we may argue back to the greater probability of the ketonic structure of the cellulose. We must note, however, that the celluloses of the lignocelluloses are obtained as residues of various reactions, and are not homogeneous. They yield on boiling with condensing acids from 6 to 9 p.ct. furfural. It is usual to regard furfural as invariably produced from a pentose residue. But this interpretation ignores a number of other probable sources of the aldehyde. It must be particularly remembered that lævulose is readily condensed (*a*) to a methylhydroxyfurfural

$$C_6H_1O_6 - 3H_2O = C_6H_6O_3 = C_5(OH).H_2.(CH_3)O_2$$

and (*b*) by HBr, with further loss of OH, as under:

$$C_6H_{12}O_6 - 4H_2O + HBr = C_5H_3(CH_2Br)O$$

and generally the ketoses are distinguished from the aldoses by their susceptibility to condensation. Such condensation of lævulose has been effected by two methods: (*a*) by heating the concentrated aqueous solution with a small proportion of oxalic acid at 3 atm. pressure [Kiermayer, Chem. Ztg. 19, 100]; (*b*) by the action of hydrobromic acid (gas) in presence of anhydrous ether; the actual compound obtained being the ω-brommethyl derivative [Fenton, J. Chem. Soc. 1899, 423].

This latter method is being extended to the investigation of typical celluloses, and the results appear to confirm the view that cellulose may be of ketonic constitution.

The evidence which is obtainable from the synthetical side of the question rests of course mainly upon the physiological [Pg 85] basis. There are two points which may be noted. Since the researches of Brown and Morris (J. Chem. Soc. 1893, 604) have altered our views of the relationships of starch and cane sugar to the assimilation process, and have placed the latter in the position of a primary product with starch as a species of overflow and reserve product, it appears that lævulose must play an important part in the elaboration of cellulose. Moreover, A. J. Brown, in studying the cellulosic cell-collecting envelope produced by the *Bacterium xylinum*, found that the proportion of this product to the carbohydrate disappearing under the action of the ferment was highest in the case of lævulose. These facts being also taken into consideration there is a concurrence of suggestion that the typical CO group in the celluloses is of ketonic character. That the typical cotton cellulose breaks down finally under the action of sulphuric acid to dextrose cannot be held to prove the aldehydic position of the carbonyls in the unit groups of the actual cellulose molecule or aggregate.

We again are confronted with the problem of the aggregate and as to how far it may affect the constitution of the unit groups. That it modifies the functions or reactivity of the ultimate constituent groups we have seen from the study of the esters. Thus with the direct ester reactions the normal fibrous cellulose ($C_6H_{16}O_5$) yields a monoacetate, dibenzoate, and a trinitrate respectively under conditions which determine, with the simple hexoses and anhydrides, the maximum esterification, i.e. all the OH groups reacting. If the OH groups are of variable function, we should expect the CO groups *a fortiori* to be susceptible of change of function, i.e. of position within the unit groups.

But as to how far this is a problem of the constitution or phases of constitution of the unit groups or of the aggregate under reaction we have as yet no grounds to determine. [Pg 86]

The subjoined communication, appearing after the completion of the MS. of the book, and belonging to a date subsequent to the period intended to be covered, is nevertheless included by reason of its exceptional importance and special bearing on the constitutional problem above discussed.

THE ACTION OF HYDROGEN BROMINE ON CARBOHYDRATES. [4]

H. J. H. Fenton and Mildred Gostling (J. Chem. Soc., 1901, 361).

The authors have shown in a previous communication (Trans., 1898, 73, 554) that certain classes of carbohydrates when acted upon at the ordinary temperature with dry hydrogen bromide in ethereal solution give an intense and beautiful purple colour. [5] It was further shown (Trans., 1899, 75, 423) that this purple substance, when neutralised with sodium carbonate and extracted with ether, yields golden-yellow prisms of ω-brommethylfurfural,

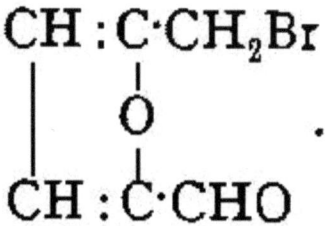

This reaction is produced by lævulose, sorbose, cane sugar, and inulin, an intense colour being given within an hour or two. Dextrose, maltose, milk sugar, galactose, and the polyhydric alcohols give, if anything, only insignificant colours, and these only after long standing. The authors therefore suggested that the reaction might be employed as a means of [Pg 87] distinguishing these classes of carbohydrates, the rapid production of the purple colour being indicative of *ketohexoses*, or of substances which produce these by hydrolysis.

By relying only on the production of the purple colour, however, a mistake might possibly arise, owing to the fact that *xylose* gives a somewhat similar colour after standing for a few hours. Hence, the observations should be confirmed by isolation of the crystals of brommethylfurfural. No trace of this substance is obtained from the xylose product.

In order to identify the substance, the ether extract, after neutralisation, is allowed to evaporate to a syrup, and crystallisation promoted either by rubbing with a glass rod, or by the more certain and highly characteristic method of 'sowing' with the most minute trace of ω-brommethylfurfural, when crystals are almost instantly formed. These are recrystallised from ether, or a mixture of ether and light petroleum, and further identified by the melting-point (59.5-60.5°), and, if considered desirable, by estimation of the bromine.

It is now found, so reactive is the bromine atom in this compound, that the estimation may be accurately made by titration with silver nitrate according to Volhard's process, the crystals for this purpose being dissolved in dilute alcohol:

0.1970 gram required 10.5 c.c. $N/10$ $AgNO_3$. Br = 42.63 p.ct., calculated 42.32 p.ct.

This method of applying hydrogen bromide in ethereal solution is, of course, unsuitable for investigations where a higher temperature has to be employed, or where long standing is necessary, since, under such circumstances, the ether itself is attacked. Wishing to make investigations under these conditions, the authors have tried several solvents, and, at present, find that chloroform is best suited to the purpose. In each of the following experiments, 10 grms. of the [Pg 88] substance were covered with 250 c.c. of chloroform which had been saturated at 0° with dry hydrogen bromide. The mixture was contained in an accurately stoppered bottle, firmly secured with an iron clamp, and heated in a water-bath to about the boiling temperature for two hours. After standing for several hours, the mixture was treated with sodium carbonate (first anhydrous solid, and afterwards a few drops of strong solution), filtered, and the solution dried over calcium chloride. Most of the chloroform was then distilled off, and the remaining solution allowed to evaporate to a thick syrup in a weighed dish.

The product was then tested for ω-brommethylfurfural by 'sowing' with the most minute trace of the substance, as described above. It was then warmed on a water-oven, kept in a vacuum desiccator over solid paraffin, and the weight estimated. When necessary, the product was recrystallised from ether, and further identified by the tests mentioned. The following results were obtained:

		Weight of crude residue.	
Swedish filter paper	3.0	crystallised at once	by 'sowing.'
Ordinary cotton	3.3	"	"

Mercerised cotton	2.1	"	"
Straw cellulose [6]	2.3	"	"
Lævulose	2.2	"	"
Inulin	1.3	"	"
Potato starch	0.37	"	"
Cane sugar	0.85	"	"
Dextrose	0.33	uncrystallisable.	
Milk sugar	0.37	"	
Glycogen	0.34	"	
Galactose	0.34	"	

[Pg 89]

The products from *dextrose, milk sugar,* and *galactose* absolutely refused to crystallise even when extracted with ether and again evaporated, or by 'sowing,' stirring, &c.

The *glycogen* product deposited a very small amount of crystalline matter on standing, but the quantity was too minute for examination; moreover, it refused altogether to crystallise in contact with the aldehyde. It may fairly be stated, therefore, that these last four substances give absolutely negative results as regards the formation of ω-brommethylfurfural; if any is formed, its quantity is altogether too small to be detected.

The specimen of *starch* examined was freshly prepared from potato, and purified by digestion for twenty-four hours each with $N/10$ KOH, $N/4$ HCl, and strong alcohol; it was then washed with water and allowed to dry in the air. It will be seen that this substance gave a positive result, but that the yield was extremely small, and might yet be due to impurity. Considering the importance of the behaviour of starch, for the purpose of drawing general conclusions from these observations, it was thought advisable to make further experiments with specimens which could be relied upon, and also to investigate the behaviour of dextrin. This the authors have been enabled to do upon a series of specimens specially prepared by C. O'Sullivan, and thus described by him:

1. Rice starch, specially purified by the permanganate method.

2. Wheat starch   "   "   "

3. Oat starch, contains traces of oil, washed with dilute KOH and dilute HCl.

4. Pea starch, first crop, washed with alkali, acid (HCl), and strong alcohol.

5. Natural dextrin, D = 3.87, a_D = 194.7; K = 0.95, (c 2.628).

6. α-Dextrin, C equation purified without fermentation, 30 precipitations with alcohol (Trans., 1879, 35, 772).

[Pg 90]

The examination of these specimens was conducted on a smaller scale, but under the same conditions as before, *one gram* of the substance being treated with 12.5 c.c. of the saturated chloroform solution and heated in sealed tubes for two hours as above. The results were as follows:

Weight of crude residue.

1. Rice starch	0.046	crystallised at once	by 'sowing.'
2. Wheat starch	0.044	"	"
3. Oat starch	0.049	"	"
4. Pea starch	0.064	"	"
5. Natural dextrin	0.088	"	"
6. α-Dextrin	0.055	"	"

The results may therefore be summarised as follows:—Treated under these particular conditions all forms of cellulose give large yields of ω-brommethylfurfural, some varieties giving as much as 33 per cent. Lævulose, inulin, and cane sugar give yields varying from 22 to 8.5 per cent.; various starches give small yields (average about 4.5 per cent.); and dextrins 5 to 8 per cent., whereas dextrose, milk sugar, and galactose give, apparently, none at all.

The yields represent the solid crystalline residue; this when purified by recrystallisation gives, probably, about three-quarters of its weight of pure crystals. (In the case of dextrose, &c., the yields represent the weight of syrup.)

These numbers, however, by no means represent the maximum yields obtainable, owing to the comparatively slight solubility of hydrogen bromide in chloroform. The process was conducted in the above manner only for the sake of uniform comparison. The ether method previously described gives much larger yields; for example, 12 grms. of inulin treated with only 60 c.c. of the saturated ether gave 2.5 grms. [Pg 91] of substance. For the purpose of obtaining larger yields, other methods are being investigated.

The facts recorded above, taken in conjunction with those given in our previous communications, appear to point definitely to the following general conclusions. First, that the various forms of *cellulose* contain one or more groups or nuclei identical with that contained in *lævulose*, and that such groups constitute the main or essential part of the molecule. Secondly, that similar groupings are contained in *starches* and *dextrins*, but that the proportion of such groupings represents a relatively small part of the whole structure.

The nature of this grouping is, according to the generally accepted constitution of *lævulose*, the six-carbon chain with a ketonic group:

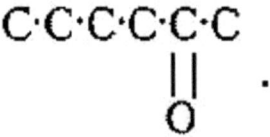

But the results might, on the other hand, be considered indicative of the anhydride or 'lacton' grouping, which Tollens suggested for lævulose:

$$C \cdot C \cdot C \cdot C \cdot C \cdot C$$
$$\diagdown \diagup$$
$$O$$

The latter very simply represents the formation of ω-brommethylfurfural from lævulose, [7]

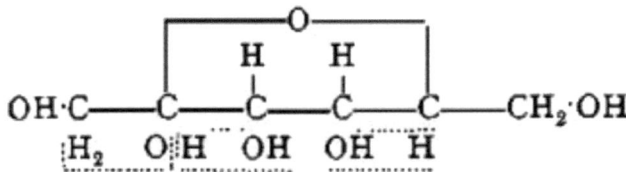

giving

$$HC \cdot C : C \cdot C : C \cdot CH_2Br$$

with H H above, and O below with \/ O ,

[Pg 92]

although by a little further 'manipulation' of the symbols the change could, of course, be represented by reference to the ketonic formula.

The Ketonic Constitution of Cellulose.

C. F. Cross and E. J. Bevan (J. Chem. Soc., 1901, 366).

In this paper the authors discuss more fully the theoretical bearings of the observations of Fenton and Gostling, the two papers being simultaneously communicated. The paper is mainly devoted to a review of the antecedent evidence, chemical and physiological, and to a general summing up in favour of the view that cellulose is a polyketose (anhydride).

(p. 79) **Composition of the Seed Hair of Eriodendron (Anf.)** — Some interest attaches to the results of an analytical investigation which we have made of this silky floss. There is little doubt that cotton is entirely exceptional in its characteristics: both in structure and chemical composition it fails to show any adaptation to what we may regard as the *more obvious* functions of a seed hair — which certainly do not demand either structural strength or chemical resistance. The following numbers determined for the kapok differentiate it widely from the cottons:

Ash, 1.3; moisture, 9.3; alkaline hydrolysis (loss) (*a*) 16.7, (*b*) 21.8. Cellulose, by chlorination, &c., 71.1.

In reacting with chloride it shows the presence of unsaturated groups, similar to the lignone of the woods. This was [Pg 93] confirmed by a well-marked reaction with ferric ferricyanide with increase of weight due to the fixation of the blue cyanide.

But the most characteristic feature is the high yield of furfural on boiling with condensing acids. The following numbers were determined:

Total furfural from original fibre	14.84
In residue from alkali hydrolysis	11.5
In cellulose isolated by Cl method	10.4

Treated with sulphuric acids of concentration, (*a*) 92.1 grs. H_2SO_4 per 100 c.c., (*b*) 105.8 grs. per 100 c.c., the fibres dissolve, and diluted immediately after complete solution it was resolved into

	(*a*)	(*b*)
Reprecipitated fraction	68.7	43.7
Soluble fraction yielding furfural	13.2	14.3

By these observations it is established that the furfuroids are of the cellulose type and behave very much as the furfuroids of the cereal celluloses.

This group of seed hairs invites exhaustive investigation. The furfuroid constituents are easily isolated, and as they constitute at least one-third of the fibre substance it is especially from this point of view that they invite study. [Pg 94]

RECHERCHES SUR L'OXYCELLULOSE.

L. Vignon.

Résumé of investigations (1898-1900) of Oxycellulose, published as a brochure (Rey, Lyon, 1900).

(*a*) A typical oxycellulose prepared from cotton cellulose by the action of $HClO_3$ (HCl + $KClO_3$) in dilute solution at 100° for one hour gave the following numbers:

	C	H	O
Elementary composition	43.55	6.03	50.42
	Oxycellulose	Original cellulose	
Analysis by Lange's method			
Soluble in KOH (at 180°)	87.6	12.0	
Insoluble in KOH (at 180°)	12.4	88.0	
	Oxycellulose	Original cellulose	
Heat of combustion	4124-4133	4190-4224	
Heat evolved in contact with 50 times wt. normal KOH per 100 grms.	1.3 cal.	0.74 cal.	

		Oxycellulose	Cellulose
Absorption of colouring matters at 100° per 100 grms.	Saffranine	0.7	0.0
	Methylene blue	0.6	0.2

(b) *Yield of furfural from cellulose, oxy- and hydro-cellulose.* — From the hydrocelluloses variously prepared the author obtains 0.8 p.ct. furfural; from bleached cotton 1.8 p.ct.; and from the oxycelluloses variously prepared 2.0-3.5 p.ct. The 'furfuroid' is relatively more soluble in alkaline solutions (KOH) in the cold. The insoluble residue is a normal cellulose.

(c) *Nitrates of cellulose, oxy- and hydro-cellulose.* — Treated with the usual acid mixture (H_2SO_4 3 p., HNO_3 1 p.) under conditions for maximum action, the resulting esters showed uniformly a fixation of [Pg 95] 11.0 NO_2 groups per unit mol. of C_{24}. The oxycellulose nitrate was treated directly with dilute solution of potassium hydrate in the cold. From the products of decomposition the author obtained the osazone of hydroxypyruvic acid [Will, Ber. 24, 400].

(d) *Osazones of the oxycelluloses.* — Oxycelluloses prepared by various methods are found to fix varying proportions of phenylhydrazine (residue), viz. from 3.4-8.5 p.ct. of the cellulose derivative reacting, corresponding with, i.e. calculated from, the nitrogen determined in the products (0.87-2.2 p.ct.). The reaction is assumed to be that of osazone formation.

The author has also established a relation between the phenylhydrazine fixed and the furfural which the substance yields on boiling with condensing acids. This is illustrated by the subjoined series of numbers:

		Fixed p.ct.	formed p.ct.
Cotton (bleached)		1.73	1.60
Oxycellulose	($HClO_3$)	7.94	2.09
"	(HClO)	3.37	1.79

"	(CrO$_3$) (1)	7.03	3.00
"	(CrO$_3$) (2)	7.71	3.09
"	(CrO$_3$) (3)	8.48	3.50

(*e*) *Constitution of cellulose and oxycellulose.* — The results of these investigations are generalised as regards cellulose (C_6) by the constitutional formula

$$(CHOH)_3 \Big\langle \begin{array}{c} CH-CH_2 \\ | \quad\quad | \\ O \\ | \\ CH-O \end{array}$$

The oxycelluloses contain the characteristic group

$$(CHOH)_3 \Big\langle \begin{array}{c} COH \\ CH-CO \\ \diagdown O \diagup \end{array}$$

in union with varying proportions of residual cellulose.

[Pg 96]

QUANTITATIVE SEPARATION OF CELLULOSE-LIKE CARBOHYDRATES IN VEGETABLE SUBSTANCES.

Wilhelm Hoffmeister (Landw. Versuchs-Stat., 1897, 48, 401-411).

To separate the hemicelluloses, celluloses, and the constituents of lignin without essential change, the substance, after being freed from fat, is extracted with dilute hydrochloric acid and ammonia, and the residue frequently agitated for a day or two with 5-6 p.ct. caustic soda solution. It is then diluted, the extract poured off, neutralised with hydrochloric acid, treated with sufficient alcohol, and the hemicellulose filtered, dried, and weighed. The residue from the soda extract is washed on a filter with hot water, and extracted with Schweizer's reagent.

When the final residue (lignin) is subjected to prolonged extraction with boiling dilute ammonia (a suitable apparatus is described, with sketch) until the ammonia is no longer coloured, a residue is obtained which mostly dissolves in Schweizer's reagent, and on repeating the process the residue is found to consist largely of mineral matter. The dissolved cellulose-like substances often contain considerable amounts of pentosanes.

According to the nature of the substance, the extraction with ammonia may take weeks, or months, or even longer; the ammonia extracts of hard woods (as lignum vitæ) and of cork are dark brown, and give an odour of vanilla when evaporated down. The residues, which are insoluble in water, but redissolve in ammonia, have the properties of humic acids. Other vegetable substances, when extracted, yielded, besides humic acids, a compound, $C_6H_7O_2$, soluble in alcohol and chloroform, but insoluble in water, ether, and benzene; preparations from different sources melted between 200° and 210°.

FOOTNOTES:

[4] The original paper is reproduced with slight alterations.

[5] This purple colour would appear to be due to a highly dissociable compound of ω-brommethylfurfural with hydrogen bromide. The aldehyde gives yellow or colourless solutions in various solvents, which are turned purple by a sufficient excess of hydrogen bromide. Dilution, or addition of water, at once discharges the colour.

[6] Other forms of cellulose were also examined — for example, pinewood cellulose — and the substances separated from solution as thiocarbonate (powder and film). All of these gave good yields of ω-brommethylfurfural.

[7] The change is empirically represented as
$C_6H_{12}O_6 + HBr - 4H_2O = C_6H_5O_2Br$.

[Pg 97]

SECTION IV. CELLULOSE GROUP, INCLUDING HEMICELLULOSES AND TISSUE CONSTITUENTS OF FUNGI

VERSUCHE ZUR BESTIMMUNG DES GEHALTS EINIGER PFLANZEN UND PFLANZENTEILE AN ZELLWANDBE-STANDTEILEN AN HEMICELLULOSEN UND AN CELLU-LOSE.

A. Kleiber (Landw. Vers.-Stat., 1900, 54, 161).

ON THE DETERMINATION OF CELL-WALL CONSTITU-ENTS, HEMICELLULOSES AND CELLULOSE IN PLANTS AND PLANT TISSUES.

In a preliminary discussion the author critically compares the results of various of the methods in practice for the isolation and estimation of cellulose. The method of F. Schulze [digestion with dil. HNO_3 with $KClO_3$ — 14 days, and afterwards treating the product with ammonia, &c.] is stated to be the 'best known' (presumably the most widely practised); W. Hoffmeister's modification of the above, in which the nitric acid is replaced by hydrochloric acid (10 p.ct. HCl) is next noted as reducing the time of digestion from 14 days to 1-2 days, and giving in many cases higher yields of cellulose. The methods of treating with the halogens, viz. bromine water (H. Müller), chlorine gas (Cross and Bevan), and chlorine water, are dismissed with a bare mention, apparently on the basis of the conclusions of Suringar and Tollens ($q.v.$). The method of Lange, the basis of which is a 'fusion' with alkaline hydrates at 180°, and the modified method of Gabriel, in which the 'fusion' with alkali takes place in presence of glycerin, are favourably mentioned.

These methods were applied to a range of widely different raw materials to determine, by critical examination of the products, both as regards yield and composition, what title these latter have to be regarded as 'pure cellulose.'

This portion of the investigation is an extension of that of Suringar and Tollens, these latter confining themselves to [Pg 98] celluloses of the 'normal' groups, i.e. textile and paper-making celluloses. The present communication is a study of the tissue and cell-wall constituents of the following types: —

1. Green plants of false oat grass (*Arrhenatherium, E.*).
2. Green plants of lucerne (*Medicago sativa*).
3. Leaves of the ash (*Fraxinus*).
4. Leaves of the walnut (*Juglans*).
5. Roots of the purple melic grass (*Molinia cærulea*).
6. Roots of dandelion (*Taraxacum officinale*).
7. Roots of comfrey.
8. Coffee berries.
9. Wheat bran.

These raw materials were treated for the quantitative estimation of cellulose by the method of Lange (*b*), Hoffmeister (*c*), and Schulze (*d*), and the numbers obtained are referred for comparison to the corresponding yields of 'crude fibre' (Rohfaser) by the standard method (*a*).

As a first result the author dismisses Lange's method as hopeless: the results in successive determinations on the same materials showing variations up to 60 p.ct. The results by *c* and *d* are satisfactorily concordant: the yields of cellulose are higher than of 'crude fibre.' This is obviously due to the conservation of 'hemicellulose' products, which are hydrolysed and dissolved in the treatments for 'crude fibre' estimation. A modified method was next investigated, in which the process of digestion with acid chloroxy- compounds (*c* and *d*) was preceded by a treatment with boiling dilute acid. The yields of cellulose by this method (*e*) are more uniform, and show less divergence from the numbers for 'crude fibre.'

The author's numerical results are given in a series of tables which include determinations of proteids and ash constituents, and the corresponding deductions from the crude [Pg 99] weight in calculating to 'pure cellulose.' The subjoined extract will illustrate these main lines of investigation.

Raw Material	Crude Fibre	Pure Cellulose	
	Weende Method. (a)	Hoffmeister Method. (c)	Hoffmeister, modified by Author. (e)
Oat grass	30.35	34.9	31.5
Lucerne	25.25	28.7	20.5
Leaves of ash	13.05	15.4	13.8
Roots of melic	21.60	29.1	21.4
Coffee beans	18.30	35.1	23.3
Bran	8.2	19.3	9.3

The final conclusion drawn from these results is that the method of Hoffmeister yields a product containing variable proportions of hemicelluloses. These are eliminated by boiling with a dilute acid (1.25 p.ct. H_2SO_4), which treatment may be carried out on the raw material—*i.e.* before exposure to the acid chlorate, or on the crude cellulose as ordinarily isolated.

Determination of Tissue-constituents.—By the regulated action of certain solvents applied in succession, it appears that such constituents of the plant-complex can be removed as have no organic connection with the cellular skeleton: the residue from such treatments, conversely, fairly represents the true tissue-constituents. The author employs the method of digestion with cold dilute alkaline solutions (0.15 to 0.5 p.ct. NaOH), followed by exhaustive washing with cold and hot water, afterwards with cold and hot alcohol, and finally with ether.

The residue is dried and weighed as crude product. When necessary, the proportions of ash and proteid constituents are determined and deducted from the 'crude [Pg 100] product' which, thus corrected, may be taken as representing the 'carbohydrate' tissue constituents.

Determination of Hemicelluloses. — By the process of boiling with dilute acids (1.25 p.ct. H_2SO_4) the hemicelluloses are attacked — i.e. hydrolysed and dissolved. The action of the acid though selective is, of course, not exclusively confined to these colloidal carbohydrates. The proteid and mineral constituents are attacked more or less, and the celluloses themselves are not entirely resistant to the action. The loss due to the latter may be neglected, but in calculating the hemicellulose constants from the gross loss the proteids and mineral constituents require to be taken into account in the usual way.

QUANTITATIVE SEPARATION OF HEMICELLULOSE, CELLULOSE, AND LIGNIN. PRESENCE OF PENTOSANES IN THESE SUBSTANCES.

Wilhelm Hoffmeister (Landw. Versuchs-Stat, 1898, 50, 347-362).

(p. 88) The separation of the cellulose-like carbohydrates of sunflower husks is described.

In order to ascertain the effect of dilute ammonia on the cellulose substances of lignin, a dried 5 p.ct. caustic soda extract was extracted successively with 1, 2, 3, and 4 p.ct. sodium hydroxide solution. Five grams of the 2 p.ct. extract were then subjected to the action of ammonia vapour; the cellulose did not completely dissolve in six weeks. Cellulose insoluble in caustic soda (32 grms.) was next extracted with ammonia, in a similar manner, for 10 days, dried, and weighed. 30.46 grms. remained, which, when treated with 5 p.ct. aqueous caustic soda, yielded 0.96 grm. (3 per cent.) of hemicellulose.

When cellulose is dissolved in Schweizer's solution, the [Pg 101] residue is, by repeated extraction with aqueous sodium hydroxide, completely converted into the soluble form. On evaporating the ammonia from the Schweizer's extract, at the ordinary temperature and on a water-bath respectively, different amounts of cellulose are obtained; more hemicellulose is obtained, by caustic soda, from the heated solution than from that which was not heated. In this operation the pentosanes are more influenced than the hexosanes; pentosanes are not always readily dissolved by caustic soda, and hexosanes are frequently more or less readily dissolved. Both occur in lignin, and are then undoubtedly indigestible. These points have to be considered in judging the digestibility of these carbohydrates.

A comparison of analyses of clover, at different periods, in the first and second years of growth, shows that both cellulose (Schweizer's extract) and lignin increase in both constituents. In the second year the lignin alone increased to the end; the cellulose decreased at the end of June. In the first year it seemed an absolutely as well as relatively greater amount of cellulose, and lignin was produced in the second year; this, however, requires confirmation. The amount of pentosanes in the Schweizer extract was relatively greater in the second than in the first year, but decreased in the lignin more in the second year than in the first: this result is also given with reserve.

DIE CONSTITUTION DER CELLULOSEN DER CEREALIEN.

C. F. Cross, E. J. Bevan, and C. Smith (Berl. Ber., 1896, 1457).

THE CONSTITUTION OF THE CEREAL CELLULOSES.

(p. 84) Straw cellulose is resolved by two methods of acid hydrolysis into a soluble furfural-yielding fraction, and an insoluble [Pg 102] fraction closely resembling the normal cellulose. (*a*) The cellulose is dissolved in sulphuric acids of concentration, $H_2SO_4.2H_2O$, $H_2SO_4.3H_2O$. As soon as solution is complete, the acid is diluted. A precipitate of cellulose hydrate (60-70 p.ct.) is obtained, and the filtered solution contains 90-95 p.ct. of the furfuroids of the original cellulose. The process is difficult to control, however, in mass, and to obtain the latter in larger quantity the cellulose (*b*) is digested with six times its weight of 1 p.ct. H_2SO_4 at 3 atm. pressure, the products of the action being (1) a disintegrated cellulose retaining only a small fraction (1/12) of the furfural-yielding groups, and (2) a slightly coloured solution of the hydrolised furfuroids. An investigation of the latter gave the following results: By oxidation with nitric acid no saccharic acid was obtained; showing the absence of dextrose. The numbers for cupric reduction were in excess of those obtained with the hexoses. The yield of ozazone was high, viz. 30 to 40 p.ct. of the weight of the carbohydrate in solution. On fractionating, the melting-points of the fractions were found to lie between 146° and 153°. Ultimate analysis gave numbers for C, H, and N identical with those of a pentosazone. The product of hydrolysis appears, therefore, to be xylose or a closely related derivative.

All attempts to obtain a crystallisation of xylose from the solution neutralised ($BaCO_3$), filtered, and evaporated, failed. The reaction with phloroglucol and HCl, moreover, was not the characteristic red of the pentoses, but a deep violet. The product was then isolated as a dry residue by evaporating further and drying at 105°. Elementary analysis gave the numbers C 44.2, 44.5, and H 6.7, 6.3. Determinations of furfural gave 39.5 to 42.5 p.ct. On treating the original solution with hydrogen peroxide, and warming, oxidation set in, with evolution of [Pg 103] CO_2. This was estimated (by absorption), giving numbers for CO_2, 19.5, 20.5, 20.1 p.ct. of the substance.

The sum of these quantitative data is inconsistent with a pentose or pentosane formula; it is more satisfactorily expressed by the empirical formula

$$C_5H_8O_3 {<}{\genfrac{}{}{0pt}{}{O}{O}}{>} CH_2,$$

which represents a pentose monoformal. Attempts to synthesise a compound of this formula have been so far without success.

UEBER EINIGE CHEMISCHE VORGÄNGE IN DER GERSTEN-PFLANZE.

C. F. Cross, E. J. Bevan, and C. Smith (Berl. Ber., 1895, 2604).

THE CHEMICAL LIFE-HISTORY OF THE BARLEY PLANT.

(p. 84) Owing to the presence of 'furfuroids' in large proportion as constituents of the tissues of the stems of cereals, these plants afford convenient material for studying the problem of the constitution of the tissue-furfuroids, as well as their relationship to the normal celluloses. The growing barley plant was investigated at successive periods of growth. Yield of furfural was estimated on the whole plant and on the residue from a treatment with alkaline and acid solvents in the cold such as to remove all cell contents. This residue is described as 'permanent tissue.' The observations were carried out through two growing seasons—1894-5—which were very different in character, the former being rainy with low temperature, the latter being abnormal in the opposite direction, i.e. minimum rainfall and maximum sunshine. The barley selected for observation was that of two experimental plots of the Royal Agricultural Society's farm, one (No. 1) remaining permanently unmanured, and showing minimum yield, the other (No. 6) receiving such fertilising treatment as to give maximum yields. [Pg 104]

The numerical results are given in the annexed tables:

BARLEY CROP, WOBURN, 1894.

Date	Age of Crop	Plot	Dry Weight	Furfural p.ct. of dry weight(a)	Permanent tissue p.ct. dry weight	Furfural from permanent tissue		Ratio $a:c$
						P.ct. of tissue	P.ct. of entire plant	

May 7	6 weeks	1	19.4	7.0	53.4	12.7	6.8	1.03 : 1
		6	14.7	7.0	55.9	10.3	5.7	1.23 : 1
June 4	10 weeks	1	17.6	7.7	52.9	11.6	6.1	1.26 : 1
		6	13.5	8.1	58.5	13.4	7.8	1.04 : 1
July 10	15 weeks	1	42.0	9.0	65.7	9.8	6.4	1.40 : 1
		6	32.9	10.6	65.7	12.5	8.2	1.30 : 1
Cut Aug. 21	21 weeks	1	64.0	11.9	70.0	14.5	10.1	1.18 : 1
		6	64.6	13.4	70.5	15.0	10.6	1.26 : 1
Carried Aug. 31	22 weeks	1	84.0	12.7	75.0	16.5	12.4	1.02 : 1
		6	86.4	12.4	78.4	15.1	11.8	1.05 : 1
BARLEY CROP, WOBURN, 1895.								
May 15	7 weeks	1	20.6	6.6	53.9	10.2	5.5	1.20 : 1
		6	17.8	5.8	56.7	9.6	5.4	1.07 : 1
June 18	12 weeks	1	34.6	8.0	38.2	14.7	5.6	1.42 : 1
		6	33.4	7.6	44.5	15.0	6.7	1.14 : 1

July 16	16 weeks	1	52.8	12.1	55.6	16.3	9.1	1.33 : 1
		6	54.4	10.6	46.2	19.1	8.8	1.20 : 1
Aug. 16	20 weeks	1	66.8	9.2	49.1	17.0	8.3	1.10 : 1
		6	65.0	9.8	49.8	19.1	9.4	1.04 : 1
Sept. 3	22 weeks	1	84.3	10.4	45.7	17.6	8.0	1.31 : 1
		6	86.3	10.2	45.3	17.3	7.8	1.30 : 1

[Pg 105]

The variations exhibited by these numbers are significant. It is clear, on the other hand, that the assimilation of the furfuroids does not vary in any important way with variations in conditions of atmosphere and soil nutrition. They are essentially *tissue*-constituents, and only at the flowering period is there any accumulation of these compounds in the alkali-soluble form. It has been previously shown (*ibid.* 27, 1061) that the proportion of furfuroids in the straw-celluloses of the paper-maker differs but little from that of the original straws. For the isolation of the celluloses the straws are treated by a severe process of alkaline hydrolysis, to which, therefore, the furfuroid groups offer equal resistance with the normal hexose groups with which they are associated in the complex.

The furfuroids of the cereal straws are therefore not pentosanes. They are original products of assimilation, and not subject to secondary changes after elaboration such as to alter either their constitution or their relationship to the normal hexose groups of the tissue-complex.

(1) CONSTITUTION OF THE CEREAL CELLULOSES

(Chem. Soc. J. 1896, 804).

(2) THE CARBOHYDRATES OF BARLEY STRAW

(Chem. Soc. J. 1896, 1604).

(3) THE CARBOHYDRATES OF THE CEREAL STRAWS (Chem. Soc. J. 1897, 1001).

(4) THE CARBOHYDRATES OF BARLEY STRAW

(Chem. Soc. J. 1898, 459).

C. F. cross, E. J. Bevan, and Claud Smith.

These are a series of investigations mainly devoted to establishing the identity of the furfural-yielding group which is a characteristic constituent.

This 'furfuroid' while equally resistant to alkalis as the normal cellulose group with which it is associated, is selectively [Pg 106] hydrolysed by acids. Thus straw cellulose dissolves in sulphuric acids of concentration $H_2SO_4.2H_2O$ - $H_2SO_4.3H_2O$, and on diluting the normal cellulose is precipitated as a hydrate, and the furfuroid remains in solution. But this sharp separation is difficult to control in mass. By heating with a very dilute acid (1 p.ct. H_2SO_4) the conditions are more easily controlled, the most satisfactory results being obtained with 15 mins. heating at 3 atm. pressure.

(1) Operating in this way upon brewers' grains the furfuroid was obtainable as the chief constituent of a solution for which the fol-

lowing experimental numbers were determined:—Total dissolved solids, 28.0 p.ct. of original 'grains'; furfural, 39.5 p.ct. of total dissolved solids, as compared with 12.5 p.ct. of total original grains; cupric reduction (calc. to total solids), 110 (dextrose = 100) osazone; yield in 3 p.ct. solution, 35 p.ct. of weight of total solids.

				Pentosazone
Analysis	N	17.1	17.3	17.07
	C	62.5	62.3	62.2
	H	6.4	6.5	6.1
Melting-point				146°-153°

From these numbers it is seen that of the total furfuroids of the original 'grains' 84 p.ct. are thus obtained in solution in the fully hydrolysed form, which is that of a pentose or pentose derivative. It was, however, found impossible to obtain any crystallisation from the neutralised ($BaCO_3$) and concentrated solution, the syrup being kept for some weeks in a desiccator. It was noted at the same time that the colour reaction of the original solution with phloroglucol and hydrochloric acid was a deep violet, in contradistinction to the characteristic red of the pentoses. On oxidation with hydrogen peroxide, in the proportion of 1 mol. H_2O_2 to 1 mol. of the carbohydrate in solution, carbonic anhydride was formed in quantity = 20.0 p.ct. of the latter. [Pg 107]

Fermentation (yeast) experiments also showed a divergence from the resistant behaviour of the pentoses, a considerable proportion of the furfuroid disappearing in a normal fermentation.

(2) The quantitative methods above described were employed in investigating the barley plant at different stages of its growth. The green plant was extracted with alcohol, the residue freed from alcohol and subjected to acid hydrolysis.

The hydrolysed extract was neutralised and fermented. In the early stages of growth the furfuroids were completely fermented, i.e. disappeared in the fermentation. In the later stages this proportion fell to 50 p.ct. In the earlier stages, moreover, the normal hexose constituents of the permanent tissue were hydrolysed in large proportion by the acid, whereas in the matured straw the hydrolysis is chiefly confined to the furfuroids. In the early stages also the permanent tissue yields an extract with relatively low cupric reduction, showing that the carbohydrates are dissolved by the acid in a more complex molecular condition.

These observations confirm the view that the furfuroids take origin in a hexose-pentose series of transformations. The proportion of furfuroid groups to total carbohydrates varies but little, viz. from 1/3 in the early stages to a maximum of 1/4 at the flowering period. At this period the differentiation of the groups begins to be marked.

Taking all the facts of (1) and (2), they are not inconsistent with the hypothesis of an internal transformation of a hexose to a pentose-monoformal. Such a change of position and function of oxygen from OH to CO within the group $-CH.OH-$ is a species of internal oxidation which reverses the reduction of formaldehyde groups in synthesising to sugars, and appears therefore of probable occurrence.

These constitutional problems are followed up in (3) by the [Pg 108] indirect method of differentiating the relationships of these furfuroids to yeast fermentation, from those of the pentoses. Straw and esparto celluloses are subjected to the processes of acid hydrolysis, and the neutralised extracts fermented. With high furfural numbers indicating that the furfuroids are the chief constituents of the extract, there is an active fermentation with production of alcohol. The cupric reduction falls in greater ratio to the original (unfermented) than the furfural. Observations on the pure pentoses — xylose and arabinose added to dextrose solutions, and then exposed to yeast action — show that in a vigorous fermentation not unduly prolonged the pentoses are unaffected, but that they do come within the influence of the yeast-cell when the latter is in a less vigorous condition, and when the hexoses are not present in relatively large proportion.

(4) The observations on the growing plant were resumed with the view of artificially increasing the differentiation of the two main groups of carbohydrates. From a portion of a barley crop the inflorescence was removed as soon as it appeared. The crop was allowed to mature, and a full comparison instituted between the products of

normal and abnormal growth. With a considerable difference in 'permanent tissue' (13 p.ct. less) and a still greater defect in cellulose (24 p.ct.), the constants for the furfuroids in relation to total carbohydrates were unaffected by the arrested development. This was also true of the behaviour of the hydrolysed extracts (acid processes) to yeast fermentation.

(5) The extract obtained from the brewers' grains by the process described in (2) was investigated in relation to animal digestion. It has been now generally established that the furfuroids as constituents of fodder plants are digested and assimilated in large proportion in passing through animal digestive tracts, and in this respect behave differently from the pentoses. The furfuroids being obtained, as described, in [Pg 109] a fully hydrolysed condition (monoses) the digestion problem presented itself in a new aspect, and was therefore attacked.

The result of the comparative feeding experiments upon rabbits was to show that in this previously hydrolysed form the furfuroids are almost entirely digested and assimilated, no pentoses, moreover, appearing in the urine.

Generally we may sum up the present solution of the problem of the relationship of the furfuroids to plant assimilation and growth as follows: — The pentoses are not produced as such in the process of assimilation; but furfural-yielding carbohydrates are produced directly and in approximately constant ratio to the total carbohydrates; they are mainly located in the permanent tissue; in the secondary changes of dehydration, &c., accompanying maturation they undergo such differentiation that they become readily separable by processes of acid hydrolysis from the more resistant normal celluloses; but in relation to alkaline treatments they maintain their intimate union with the latter. They are finally converted into pentoses by artificial treatments, and into pentosanes in the plant, with loss of 1 C atom in an oxidised form. The mechanism of this transformation of hexoses into pentoses is not cleared up. It is independent of external conditions, *e.g.* fertilisation and atmospheric oxidations, and is probably therefore a process of internal rearrangement of the character of an oxidation.

ZUR KENNTNISS DER IN DEN MEMBRANEN DER PILZE ENTHALTENEN BESTANDTHEILE.

E. Winterstein (Ztschr. Physiol. Chėm., 1894, 521; 1895, 134).

ON THE CONSTITUENTS OF THE TISSUE OF FUNGI.

(p. 87) These two communications are a contribution of fundamental importance, and may be regarded as placing the [Pg 110] question of the composition of the celluloses of these lowest types on a basis of well-defined fact. In the first place the author gives an exhaustive bibliography, beginning with the researches of Braconnot (1811), who regarded the cellular tissue of these organisms as a specialised substance, which he termed 'fungin.' Payen rejects this view, and regards the tissue, fully purified by the action of solvents, as a cellulose ($C_6H_{10}O_5$). This view is successively supported by Fromberg [Mulder, Allg. Phys. Chem., Braunschweig, 1851], Schlossberger and Doepping [Annalen, 52, 106], and Kaiser. De Bary, on a review of the evidence, adopts this view, but, as the purified substance fails to give the characteristic colour-reactions with iodine, he uses the qualifying term 'pilzcellulose' [Morph. u. Biol. d. Pilze u. Flechten, Leipzig, 1884].

C. Richter, on the other hand, shows that these reactions are merely a question of methods of purification or preparation [Sitzungsber. Acad. Wien, 82, 1, 494], and considers that the tissue-substance is an ordinary cellulose, with the ordinary reactions masked by the presence of impurities. In regard to the lower types of fungoid growth, such as yeast, the results of investigators are more at variance. The researches of Salkowski (p. 113) leave little doubt, however, that the cell-membrane is of the cellulosic type.

The author's researches extend over a typical range of products obtained from *Boletus edulis, Agaricus campestris, Cantharellus cibarius, Morchella esculenta, Polyporus officinalis, Penicillium glaucum*, and certain undetermined species. The method of purification consisted mainly in (*a*) exhaustive treatments with ether and boiling alcohol, (*b*) digestion with alkaline hydrate (1-2 p.ct. NaOH) in the cold, (*c*) acid hydrolysis (2-3 p.ct. H_2SO_4) at 95°-100°, followed by a chloroxidation treatment by the processes of Schulze or Hoffmeister, and final alkaline hydrolysis. [Pg 111]

The products, i.e. residues, thus obtained were different in essential points from the celluloses isolated from the tissues of phanerogams similarly treated. Only in exceptional cases do they give blue reactions with iodine in presence of zinc chloride or sulphuric acid. The colourations are brown to red. They resist the action of cuprammonium solutions. They are for the most part soluble in alkaline hydrate solution (5-10 p.ct. NaOH) in the cold. They give small yields (1-2 p.ct.) of furfural on boiling with 10 p.ct. HCl.Aq.

Elementary analyses gave the following results, which are important in establishing the presence of a notable proportion of nitrogen, which has certainly been overlooked by the earlier observers: —

'Cellulose' or residue from	C	H	N
Boletus edulis (Schulze process)	42.4	6.5	3.9
Boletus edulis (Hoffmeister process)	44.6	6.3	3.6
Polyporus off.	43.7	6.5	0.7
Cantharellus cib.	44.9	6.8	3.0
Agaricus campestris	44.3	6.6	3.6
Botrytis	42.1	6.3	3.9
Penicillium glaucum	—	—	3.3
Morchella esculenta	—	—	2.5

It is next shown that this residual nitrogen is not in the form of residual proteids (1) by direct tests, all of which gave negative results, and (2) indirectly by the high degree of resistance to both alkaline and acid hydrolysis. The 'celluloses' are attacked by boiling dilute acids (1 p.ct. H_2SO_4), losing in weight from 10 to 23 p.ct., the dissolved products having a cupric reduction value about 50 p.ct. that of an equal weight of dextrose. As an extreme hydrolytic treatment the products were dissolved in 70 p.ct. H_2SO_4, allowed to stand 24 hours, then considerably diluted (to 3 p.ct. H_2SO_4) and boiled to [Pg 112] complete the inversion. The yields of glucose, calculated from the cupric reduction, were as follows: —

Boletus edulis	65.2	p.ct.
Polyporus off.	94.7	"
Agaricus campestris	59.1	"
Morchella esculenta	60.1	"
Cantharellus cib.	64.9	"
Botrytis	60.8	"

It will be noted that the exceptionally high yield from the Polyporus cellulose is correlated with its exceptionally low nitrogen. By actual isolation of a crystalline dextrorotary sugar, by preparations of osazone and conversion into saccharic acid, it was proved that dextrose was the main product of hydrolysis. The second main product was shown to be acetic acid, the yield of which amounted to 8 p.ct. in several cases.

Generally, therefore, it is proved that the more resistant tissue constituents of the fungi are not cellulose, but a complex of carbohydrates and nitrogenous groups in combination, the former being resolved into glucoses by acid hydrolysis, and the latter yielding acetic acid as a characteristic product of resolution together with the nitrogenous groups in the form of an uncrystallisable syrup.

In the further prosecution of these investigations (2) the author proceeded from the supposition of the identity of the nitrogenous complex of the original with chitin, and adopted the method of Ledderhose (Ztschr. Physiol. Chem. 2, 213) for the isolation of glucosamin hydrochloride, which he succeeded in obtaining in the crystalline form. In the meantime E. Gilson had shown that these tissue substances in 'fusion' with alkaline hydrates yield a residue of a nitrogenous product ($C_{14}H_{28}N_2O_{10}$), which is soluble in dilute acids [Recherches Chim. sur la Membrane Cellulaire des Champignons, [Pg 113] La Cellule, v. II, pt. 1]. This residue, which was termed mycosin by Gilson, has been similarly isolated by the author. It is proved, therefore, that the tissues of the fungi do contain a product resembling chitin. [See also Gilson, Compt. Rend. 120, 1000.] This constituent is in intimate union with the carbohydrate complex, which is resolved similarly to the hemicelluloses. Various intermediate terms of the hydrolytic series have been isolated. But the only fully identified product of resolution is the dextrose which finally results.

UEBER DIE KOHLENHYDRATE D. HEFE.

E. Salkowski (Berl. Ber., 27, 3325).

ON THE CARBOHYDRATES OF YEAST.

The author has isolated the more resistant constituents of the cell-membrane by boiling with dilute alkalis, and exhaustively purifying with alcohol and ether.

The residue was only a small percentage (3-4 p.ct) of the original, and retained only 0.45 p.ct. N.

It was heated in a digester with water at 2-3 atm. steam-pressure, and thus resolved into approximately equal portions of soluble cellulose (*a*) and insoluble (*b*). The latter, giving no colour-reaction with iodine, is termed achroocellulose; the former reacts, and is therefore termed erythrocellulose. The former is easily separated from its opalescent solution. It has the empirical composition of cellulose. In the soluble form it resembles glycogen. The achroocellulose is isolated in the form of horny or agglomerated masses. It appears to be resolved by ultimate hydrolysis into dextrose and mannose.

[Pg 114]

SECTION V. FURFUROIDS, *i.e.* PENTOSANES AND FURFURAL-YIELDING CONSTITUENTS GENERALLY

(1) Reactions of the Carbohydrates with Hydrogen Peroxide.

C. F. Cross, E. J. Bevan, and Claud Smith (J. Chem. Soc., 1898, 463).

(2) Action of Hydrogen Peroxide on Carbohydrates in the Presence of Ferrous Salts.

R. S. Morrell and J. M. Crofts (J. Chem. Soc., 1899, 786).

(3) Oxidation of Furfuraldehyde by Hydrogen Peroxide.

C. F. Cross, E. J. Bevan, and T. Heiberg (J. Ch. Soc., 1899, 747).

(4) EINWIRKUNG VON WASSERSTOFFHYPEROXID AUF UNGESÄTTIGTE KOHLENWASSERSTOFFE.

C. F. Cross, E. J. Bevan, and T. Heiberg (Berl. Ber., 1900, 2015).

ACTION OF HYDROGEN PEROXIDE ON UNSATURATED HYDROCARBONS.

The above series of researches grew out of the observations incidental to the use of the peroxide on an oxidising agent in investigating the hydrolysed furfuroids (102). Certain remarkable observations had previously been made by H. J. H. Fenton (Ch. Soc. J., 1894,

899; 1895, 774; 1896, 546) on the oxidation of tartaric acid by the peroxide, acting in presence of ferrous salts, the —CHOH—CHOH— residue losing H_2 with production of the unsaturated group,
—OH.C=C.OH—. These investigations have subsequently been considerably developed and generalised by Fenton, but as the results have no [Pg 115] immediate bearing on our main subject we must refer readers to the J. Chem. Soc., 1896-1900.

From the mode of action diagnosed by Fenton it was to be expected that the CHOH groups of the carbohydrates would be oxidised to CO groups, and it has been established by the above investigations (1) and (2) that the particular group to be so affected in the hexoses is that contiguous to the typical

|
—CO

group. There results, therefore, a dicarbonyl derivative ('osone'), which reacts directly with 2 mol. phenyl hydrazine in the cold to form an osazone. This was directly established for glucose, lævulose, galactose, and arabinose (2). While this is the main result, the general study of the product shows that the oxidation is not simple nor in direct quantitative relationship to the H_2O_2 employed. The molecular proportion of the aldoses affected appears to be in considerable excess, and the reaction is probably complicated by interior rearrangement.

In the main, the original aldehydic group resists the oxidation. But a certain proportion of acid products are formed, probably tartronic acid. On distillation with condensing acids a large proportion of volatile monobasic acids (chiefly formic) are obtained. The proportion of furfural obtained amounts to 3-4 per cent. of the weight of the original carbohydrate.

Since the general result of these oxidations is the substitution of an OH group for an H atom, it was of interest to determine the behaviour of furfural with the peroxide. The oxidation was carried out in dilute aqueous solution of the aldehyde at 20°-40°, using 2-3 mols. H_2O_2 per 1 mol. $C_5H_4O_2$. The main product is a hydroxyfurfural, which was separated as a hydrazone. A small quantity of a monobasic acid was formed, which was identified as a hydroxypyromucic acid. Both aldehyde and acid appear to be the α β derivatives. The [Pg 116] aldehyde gives very characteristic colour reactions with phloroglucinol and resorcinol in presence of hydrochloric acid, which so closely resemble those of the lignocelluloses that there is little doubt that these particular reactions must be referred to the presence of the hydroxyfurfural as a normal constituent.

The study of these oxidations was then extended to typical unsaturated hydrocarbons — viz. acetylene and benzene. (4) From the former the main product was acetic acid, but the attendant formation of traces of ethyl alcohol indicates that the hydrogen of the peroxide may take a direct part in this and other reactions. This view receives some support from the fact that the interaction of the H_2O_2 with permanganates has now been established to be an oxidation of the H_2 of the peroxide by the permanganate oxidation, with liberation, therefore, of the O_2 of the peroxide as an unresolved molecule [Baeyer].

Benzene itself is also powerfully attacked by the peroxide when shaken with a dilute solution in presence of iron salts. The products are phenol and pyrocatechol, with some quantity of an amorphous product probably formed by condensation of a quinone with the phenolic products of reaction.

These types of oxidation effects now established give a definite significance to the physiological functions of the peroxide, which is a form of 'active oxygen' of extremely wide distribution. It would have been difficult *a priori* to devise an oxidant without sensible action on aldehydic groups, yet delivering a powerful attack on hydrocarbon rings; or to have suggested a synthesis of the sugars from tartaric acid with a powerful oxidising treatment as the first and essential stage in the transformation.

Our present knowledge of such actions and effects suggests [Pg 117] a number of new clues to genetic relationships of carbon compounds within the plant. The conclusion is certainly justified that the origin of the pentoses is referable to oxidations of the hexoses, in which this form of 'active oxygen' plays an important part.

We must note here the researches of O. Ruff, who has applied these oxidations with important results in the systematic investigation of the carbohydrates.

UEBER DIE VERWANDLUNG DER *D*-GLUCONSÄURE IN *D*-ARABINOSE (Berl. Ber., 1898, 1573).

CONVERSION OF *D*-GLUCONIC ACID INTO *D*-ARABINOSE.

D UND *L* ARABINOSE (*Ibid.* 1899, 550).

ZUR KENNTNISS DER OXYGLUCONSÄURE (*Ibid.* 1899, 2269).

ON OXYGLUCONIC ACID.

Ruff in these researches has realised a simple and direct transition from the hexoses to the pentoses. By oxidising gluconic acid with the peroxide the β —CHOH— group is converted into carbonyl at the same time that the terminal COOH [α] is oxidised to CO_2. The yields of the resulting pentose are large. Simultaneously there is formed an oxygluconic acid, which appears to be a ketonic acid of formula $-CH_2OH.CO.(CHOH)_3.COOH-$.

From these results we see a further range of physiological probabilities; and with the concurrent actions of oxygen in the forms of or related to hydrogen peroxide on the one side, and ozone on the other, we are able to account in a simple way for the relationships of the 'furfuroid' group, which may [Pg 118] include a number of intermediate terms in the hexose-pentose series.

Following in this direction of development of the subject is a study of the action of persulphuric acid upon furfural.

EINWIRKUNG DES CARO'SCHEN REAGENS AUF FURFURAL.

C. F. Cross, E. J. Bevan, and J. F. Briggs (Berl. Ber., 1900, 3132).

Regarding this reagent as another form of 'active oxygen,' it is important to contrast its actions with those of the hydrogen peroxide. Instead of the β-hydroxyfurfural (*ante*, 115) we obtain the δ-aldehyde as the first product. The aldehydic group is then oxidised, and as a result of attendant hydrolysis the ring is broken down and succinic acid is formed, the original aldehydic group of the furfural being split off in the form of formic acid. The reactions take place at the ordinary temperature and with the dilute form of the reagent described by Baeyer and Villiger (Ber. 32, 3625). These results have some special features of interest. The α δ-hydroxyfurfural has similar colour reactions to those of the α β-derivative, and may also therefore be present as a constituent of the lignocelluloses. The tendency to attack in the 1 4 position in relation to an aldehydic group further widens the capabilities of 'active oxygen' in the plant cell. Lastly, this is the simplest transition yet disclosed from the succinyl to furfural grouping, being effected by a regulated proportion of oxygen, and under conditions of reaction which may be described as of the mildest. In regard to the wide-reaching functions of asparagin in plant life, we have a new suggestion of genetic connections with the furfuroids. [Pg 119]

VERGLEICH DER PENTOSEN-BESTIMMUNGSMETHODEN VERMITTELST PHENYLHYDRAZIN UND PHLOROGLUCIN.

M. Krüger (Inaug.-Diss., Göttingen, 1895).

COMPARISON OF METHODS OF ESTIMATING FURFURAL AS HYDRAZONE AND PHLOROGLUCIDE.

The author traces the development of processes of estimating furfural (1) by precipitation with ammonia (furfuramide), (2) by volumetric estimation with standardised phenylhydrazine, (3) by weighing the hydrazone.

In 1893 (Chem. Ztg. 17, 1745) Hotter described a method of quantitative condensation with pyrogallol requiring a temperature of 100°-110° for two hours. The insoluble product collected, washed, dried at 103°, and weighed, gives a weight of 1.974 grm. per 1 grm. furfural.

Councler substitutes phloroglucinol for pyrogallol, with the advantage of doing away with the digestion at high temperature. (*Ibid.* 18, 966.) This process, requiring the presence of strong HCl, has the advantage of being applied directly to the acid distillate, in which form furfural is obtained as a product of condensation of pentoses, &c. A comparative investigation was made, precipitating furfural (*a*) as hydrazone in presence of acetic acid, and (*b*) as phloroglucide in presence of HCl (12 p.ct). In (*a*) by varying the weights of known quantities of furfural, and using the factor, hydrazone × 0.516 [+ 0.0104] in calculating from the weights of precipitates obtained, the maximum variations from the theoretical number were +1.71 and -1.74. In (*b*) it was found necessary to vary the factor from 0.52 to 0.55 in calculating from phloroglucide to furfural. The greatest *total* range of variation was found to be 2.5 p.ct. The phenol process is therefore equally accurate, has the advantages above noted, and, in [Pg 120] addition, is less liable to error from the pressure in the distillates obtained from vegetable substances of volatile products, e.g. ketonic compounds, accompanying the furfural.

This method has been criticised by Helbel and Zeisel [Sitz.-ber, Wiener Akad. 1895, 104, ii. p. 335] on two grounds of error, viz. (1) the presence of diresorcinol in all ordinary preparations of phloroglucinol, and (2) changes in weight of the precipitate of phloroglucide on drying. The process was carried out comparatively with ordinary preparations, and with specially pure preparations of the phenol. The quantitative results were identical. The criticisms in question are therefore dismissed. Although the process is to be recommended for its simplicity and the satisfactory concordance of results it is to be noted that it rests upon an empirical basis, since the phloroglucide is not formed by the simple reaction $2\ [C_5H_4O_2 + C_6H_6O_3] - H_2O = C_{22}H_{18}O_9$, but appears to have the composition $C_{16}H_{12}O_6$.

In part ii. of this paper the author discusses the question of the probable extent in the sense of diversity of constitution of furfural-yielding constituents of plant-tissues. Glucoson was isolated from glucosazon, and found to yield 2.9-3.6 p.ct. furfural. Gluconic acid distilled with hydrochloric acid gave traces of furfural; so also with sulphuric acid and manganic oxide.

Starch was oxidised with permanganate, and a mixture of products obtained of which one gave a characteristic violet colouration with phloroglucol, with an absorption-band at the D line. On distilling with HCl furfural was obtained in some quantity. The product in question was found to be very sensitive to the action of bases, and was destroyed by the incidental operation of neutralising the mixture of oxidised products with calcium carbonate. It was found impossible to isolate the compound. [Pg 121]

UNTERSUCHUNGEN UEBER DIE PENTOSANBESTIMMUNG MITTELST DER SALZSÄURE-PHLORO-GLUCIN-METHODE. [8]

E. Kröber (Journ. f. Landwirthschaft, 1901, 357).

INVESTIGATION OF THE HYDROCHLORIC ACID-PHLOROGLUCINOL METHOD OF DETERMINING PENTOSANES.

This paper is the most complete investigation yet published of the now well-known method of precipitating and estimating furfural in acid solution by means of the trihydric phenol. In the last section of the paper is contained the most important result, the proof that the insoluble phloroglucide is formed according to the reaction

$C_5H_4O_2 + C_6H_6O_3 - 2H_2O = C_{11}H_6O_3,$

also, by varying the proportions of the pure reagents interacting, that the condensation takes place invariably according to this equation.

Incidentally the following points were also established: — The solubility of the phloroglucide, under the conditions of finally separating in a condition for drying and weighing, is 1 mgr. per 100 c.c. of total solution, made up of the original acid solution, in which the precipitation takes place, and the wash-water required to purify from the acid. The phloroglucide is hygroscopic, and must be weighed out of contact with the air. The presence of diresorcinol is without influence on the result, provided a sufficient excess of actual phloroglucinol is employed. Thus even with a preparation containing 30 p.ct. of its weight of diresorcinol the influence of the latter is eliminated, provided a weight be taken equal to twice that of the furfural to be precipitated. The phenol must be perfectly dissolved by warming with dilute HCl (1.06 sp.gr.) before adding to the furfural [Pg 122] solution. For collecting the precipitate of phloroglucide the author employs the Gooch crucible.

The paper contains a large number of quantitative results in proof of the various points established, and concludes with elaborate tables, giving the equivalents in the known pentoses and their anhydrides for any given weight of phloroglucide from 0.050 to 0.300 grm.

UEBER DEN PENTOSAN-GEHALT VERSCHIEDENER MATERIALIEN.

B. Tollens and H. Glaubitz (J. für Landwirthschaft, 1897, 97).

ON THE PENTOSANE CONSTITUENTS OF FODDER-PLANTS AND MALT.

(p. 171) (*a*) The authors have re-determined the yield of furfural from a large range of plant-products, using the phloroglucol method. The numbers approximate closely to those obtained by the hydrazone method. The following may be cited as typical:

Substance	Furfural p.ct.
Rye (Göttingen)	6.03
Wheat (square head)	4.75
Barley (peacock)	4.33
Oats (Göttingen)	7.72
Maize (American)	3.17
Meadow hay	11.63

Bran (wheat)	13.06
Malt	6.07
Malt-sprouts	8.56
Sugar-beet (exhausted)	14.95

(b) A comparison of wheat with wheat bran, &c. was made by grinding in a mortar and 'bolting' the flour through a fine silk sieve. The results showed:

	Furfural p.ct.
Original wheat	4.75
Fine flour	1.73
Bran (24 p.ct. of wheat)	11.25
Wheat-bran of commerce	13.06

[Pg 123]

It is evident that the pentosanes of wheat are localised in the more resistant tissues of the grain.

(c) An investigation of the products obtained in the analytical process for 'crude fibre' gave the following:

(1) In the case of brewers' grains:

100 grms. grains	gave furfural		=	29.43 pentosane
20 " crude fibre	"		=	2.52
Acid extract	"		=	22.76
Alkali "	"		=	1.20
Deficiency from	total of original grains			2.95
				29.43

(2) In the case of meadow hay:

The crude fibre (30 p.ct.) obtained retained about one fourth (23.63 p.ct.) of the total original pentosanes.

(*d*) An investigation of barley-malt, malt-extract or wort, and finished beer showed the following: An increase of furfuroids in the process of malting, 100 pts. barley with 7.97 of 'pentosane' yielding 82 of malt with 11.18 p.ct. 'pentosane'; confirming the observations of Cross and Bevan (Ber. 28, 2604). Of the total furfuroids of malt about 1/4 are dissolved in the mashing process. In a fermentation for lager beer it was found that about /10 of the total furfuroids of the malt finally survive in the beer; the yield of furfural being 2.92 p.ct. of the 'total solids' of the beer. In a 'Schlempe' or 'pot ale,' from a distillery using to 1 part malt 4 parts raw grain (rye), yield of furfural was 9 p.ct. of the total solids.

In a general review of the relationships of this group of plant-products it is pointed out that they are largely digested by animals, and probably have an equal 'assimilation' value to starch. They resist alcoholic fermentation, and must consequently be taken into account as constituents of beers and wines. [Pg 124]

UEBER DAS VERHALTEN DER PENTOSANE DER SAMEN BEIM KEIMEN. [9]

A. Schöne and B. Tollens (Jour. f. Landwirthschaft, 1901, 349).

BEHAVIOUR OF PENTOSANES OF SEEDS IN GERMINATION.

The authors have investigated the germination of barley, wheat, and peas, in absence of light, and generally with exclusion of assimilating activity, to determine whether the oxidation with attendant loss of weight, which is the main chemical feature of the germination proper, affects the pentosanes of the seeds. The following are typical of the quantitative results obtained, which are stated in absolute weights, and not percentages.

--	Original seed	Malt or germinated product	Pentosane in

	A	B	A	B
Barley	500.00	434.88	39.58	40.38
"	500.00	442.26	40.52	41.17
Peas	300.00	286.60	15.25	15.97

The authors conclude generally that there is a slight absolute increase in the pentosanes, and that the pentosanes do not belong to those reserve materials which undergo destructive oxidation during germination.

In this they confirm the previously published results of De Chalmot, Cross and Bevan, and Gotze and Pfeiffer.

UEBER DEN GEHALT DER BAUMWOLLE AN PENTOSAN.

H. Suringar and B. Tollens (Ztschr. angew. Chem., 1897, I).

PENTOSANE CONSTITUENTS OF COTTON.

(p. 290) It has been stated by Link and Voswinkel (Pharm. Centralhalle, 1893, 253), that raw cotton yields [Pg 125] 'wood gum' as a product of hydrolysis. The authors were unable to obtain any pentoses as products of acid hydrolysis of raw cotton, and traces only of furfural-yielding carbohydrates. They conclude that raw cotton contains no appreciable quantity of pentosane.

FOOTNOTES:

[8] This paper appears during the printing of the author's original MS.

[9] This paper appears during the printing of the author's original MS.

SECTION VI. THE LIGNOCELLULOSES

(p. 131) **Lignocellulose Esters.**—By a fuller study of the ester reactions of the normal celluloses we have been able to throw some light on the constitutional problems involved; and we have extended the investigations to the jute fibre as a type of the lignocelluloses, from the results of which we get a clearer idea of the relationships of the constituent groups.

Taking the empirical expression for the complex, i.e. the entire lignocellulose, the formula $C_{12}H_{18}O_9$, we shall be able to compare the ester derivatives with those of the celluloses, which we have also referred to a C_{12} unit. But we shall require also to deal with the constituent groups of the complex, which for the purposes of this discussion may be regarded as (*a*) a cellulose of normal characteristics—cellulose α; (*b*) a cellulose yielding furfural on boiling with condensing acids—cellulose β; and (*c*) a much condensed, and in part benzenoid, group which we may continue to term the lig*none* group.

The latter has been specially examined with regard to its proportion of OH groups, as a necessary preliminary to the investigation of esters, in producing which the entire complex is employed. It will be shown that the ester groups can be actually localised in various ways, as in the main entering the cellulose residues α and β. But that the lignone group takes little part in the reactions may be generally concluded on the evidence of its non-reactivity as an isolated derivative, (1) By [Pg 126] chlorination, &c. it is isolated in the form of an amorphous body, but of constant composition, represented by the formula $C_{19}H_{18}Cl_4O_9$. This compound, soluble in acetic anhydride, was boiled with it for six hours after adding fused sodium acetate, and the product separated by pouring into water. The dilute acid filtered from the product contained no hydrochloric acid nor by-products of action. The product showed an increase of weight of 7.5 p.ct. For one acetyl per 1 mol. $C_{19}H_{18}Cl_4O$ the calculated increase is 8.0 p.ct. It is evident from the nature of the derivative that this result cannot be further verified by the usual analytical methods. (2) The chlorinated derivative is entirely soluble in sodium sulphite solution. This solution, shaken with benzoyl chloride, with addition of sodium hydrate in successive portions, shows only a small formation of insoluble benzoate, which separates as a tarry precipitate. (3) The empirical formula of the lignone complex in its isolated forms indicates that very little hydrolysis occurs in the processes of isolation. Thus the chlorinated product we may assume to be derived from the complex $C_{19}H_{22}O_9$. In the soluble by-products from the bisulphite processes of pulping wood the lignone exists as a sulphonated derivative, $C_{24}H_{23}(OCH_3)_2.(SO_3H).O_7$. The original lignone may be regarded as passing into solution as a still condensed complex derived from $C_{24}H_{26}O_{12}$ (Tollens). There is evidently little attendant hydroxylation, and another essential feature is the small molecular proportion of groups showing the typical sulphonation.

It appears that in the lignone the elements are approximately in the relation C_6: H_6: O_3, and it may assist this discussion to formulate the main constitutional types consistent with this ratio, viz.:

(1) The trihydroxybenzenes $C_6H_3(OH)_3$.
[Pg 127] (2) Methylhydroxyfurfural $C_5H_2O.(OH)(CH_3)$.
(3) Methylhydroxypyrone

$$CO < C_4H_2 \begin{matrix} (CH_3) \\ \\ (OH) \end{matrix} > O.$$

(4) Trioxycyclohexane

$$\begin{matrix} | & & & & & | \\ CH-CH-CH-CH-CH-CH \\ \diagdown O \diagup & \diagdown O \diagup & \diagdown O \diagup \end{matrix}$$

It is probable that all these types of condensation are represented in the lignone molecules, since the derivatives yielded in decompositions of more or less regulated character are either directly derived from or related to such groups. For the moment we pass over all but the general fact of complexity and the marked paucity of OH-groups. It would be of importance to be able to formulate the exact mode of union of the lignone with the cellulose residues to constitute the lignocellulose. The evidence, however, does not carry us farther than the probability of union by complicated groups and of large dimensions; for not only is the lignone isolated in condensed and non-hydroxylated forms, but the cellulose also is not hydrated or hydrolysed further than in the ratio $3C_6H_{10}O_5.H_2O$. It is probable, therefore, that the water combining with the residues at the moment of their resolution is relatively small.

Lastly, we have to remember, when dealing with the statistical results of the reactions to be described, that the approximate proportions per cent. of the constituent groups are:

Cellulose	α	65	}	
"	β	15	} =	100 lignocellulose.
Lignone		20	}	

Jute Benzoates. — In preparing the jute for treatment it was boiled in alkaline solution (1 per cent. NaOH), washed with water and dilute acid, again washed, dried, and weighed.

In the ester reaction the reagents were employed in the [Pg 128] proportion $C_{12}H_{18}O_9$: 3NaOH: $2C_6H_5COCl$. A series of quantitative experiments gave yields of 126-130 p.ct. of benzoate [calculated for monobenzoate 134 p.ct.].

The results were confirmed by ultimate analysis. The monobenzoate therefore represents a maximum, and this molecular proportion is one-half of that observed with the normal cellulose, calculated to the same unit.

Localisation of Benzoyl Group. — The entrance of the ester group affects the typical colour reactions of the lignocellulose, which are fainter. The ferric ferricyanide reaction almost disappears. The lignone group is unaffected, and combines with chlorine as in the original. The lignone chloride is removed by sodium sulphite solution, and the residue is a *cellulose benzoate*. The loss of weight due to the elimination of the lignone was 12.7 p.ct. Calculating per 100 of the original lignocellulose this becomes 16. These statistics further confirm the localisation of the benzoyl group in the cellulose residue. It is to be noted that the presence of the benzoyl group renders the cellulose more resistant to hydrolytic actions. Thus, to bring out this fact more prominently, we may calculate the yield of residual cellulose benzoate p.ct. of original jute, and we find it 109 p.ct. Taking a maximum proportion for original cellulose—viz. 85—this benzoate represents a yield of 129 p.ct., as against the theoretical for a monobenzoate, 132 p.ct.

Furfural Numbers.—The percentage of furfural obtained by boiling with HCl of 1.06 sp.gr. was 3.02 and 3.29 in separate determinations. Calculating to the original lignocellulose, the percentage, 4.21, indicates a considerable loss of the furfural-yielding constituent. The effect was also apparent in the cellulose (benzoate) isolated by chlorination &c., the percentage being 1.39 p.ct., and calculated to the original jute benzoate 1.59 p.ct. Under the conditions adopted in dissolving [Pg 129] away the chlorinated lignone the original non-benzoated lignocellulose would have yielded a cellulose giving 6 to 7 p.ct. furfural.

Since the benzoyl group is hardly calculated to produce a constitutional change affecting the furfural constants, it was necessary to examine the effect of the preliminary alkaline treatment, and the change in the furfuroid group was in fact localised in this reaction. It was found that, on washing the alkali from the mercerised jute, and further purifying the residue, this latter yielded only 4.2 p.ct. furfural [3.4 p.ct. on original fibre]. The alkaline solution and washings were acidified and distilled from 10 p.ct. HCl, yielding an additional 3.6 p.ct. calculated to the original lignocellulose. By treatment with the concentrated alkali, therefore, the furfuroid of the original lignocellulose undergoes little change, but is selectively dissolved. This point is under further investigation.

(p. 132) **Acetylation of Lignocelluloses.**—Acetates are readily formed by boiling the lignocelluloses with acetic anhydride. The derivatives obtained from jute are only generally mentioned in the 1st edition (p. 132). A further study of the reactions in regard to special points has led to some more definite results. The *yields* of product by the ordinary and simple process are 114-115 p.ct. But on analysing the product an important discrepancy is revealed.

For the saponification we employ a solution of sodium ethylate in the cold. The following numbers were obtained:

Acetic acid   Hydrocellulose residue

	27.2	77.8
Calc. for diacetate on $C_{12}H_{18}O_9$	30.8	78.4

The derivative is approximately a diacetate, and on the assumption of a simple ester reaction the yield should be [Pg 130] 127 p.ct. Assuming that the difference of 13 p.ct. is due to loss of water by internal condensation, it appears that for each acetyl group entering, 2 mol. H_2O are split off.

The jute acetate showed the normal reaction with chlorine, and the lignone chloride was dissolved by treatment with sodium sulphite solution. The fibrous residue was colourless. It proved to be a cellulose acetate. The following numbers were obtained on saponification:

	Acetic acid	Cellulose
	31.6	70.0
	30.9	68.8
Calc. for diacetate on $C_{12}H_{20}O_{10}$	29.4	79.9

The interpretation of these numbers appears to be this: in the original reaction with the lignocellulose it is the cellulose residue which is acetylated, and at the same time condensed. The cellulose residue which undergoes condensation is not of the normal constitution, since the normal cellulose is acetylated without condensation (see p. 41). On saponification a portion of the cellulose, in again combining with water, is hydrolysed to soluble products. The lignone group as it exists in the lignocellulose has no free OH groups, and probably no free aldehydic groups such as would react with the anhydride. Such groups may, however, be originally present, and may take part in the internal condensations which have been shown to occur. The furfural constants of the lignocellulose are unaffected by the acetylation and condensation. The hygroscopic moisture of the product is lowered from 10-11 p.ct. in the original to 4.5 p.ct. The ferric ferricyanide reaction is inhibited by the disappearance of the reactive groups, upon which this curious and characteristic phenomenon depends (1st ed.).

[Pg 131]

Acetylation of Benzoates. — The cellulose dibenzoate (C_{12} basis) and the jute monobenzoate were acetylated under comparative conditions The results were as follows:

		C12 basis			
		Cellulose dibenzoate		Jute monobenzoate	
		Found	Calc. for diacetate on dibenzoate	Found	Calc. for diacetate on monobenzoate
Ester reaction					
Yield		111 p.ct.	115 p.ct.	124 p.ct.	120 p.ct.
Saponification	{Cellulose}	53.5	52.6	59.8	61.9
	{Lignocellulose}				
	NaOH combining	21.3	23.9	28.4	24.3

From these results it would appear that the number of acetyl groups entering the benzoates is the same as with the unbenzoylated fibres, the benzoyl has no influence upon the hydroxyls as against the acetyl. At the same time the internal condensation noticed in the acetylation of the jute appears not to occur in the case of the benzoate.

Nitric Esters. — The numbers resulting from the quantitative study of the ester reaction and product (1st ed. p. 133) show a very large divergence of the yield of product from that which would be calculated from its composition (N p.ct.) on the assumption that the ester reaction is simple. We have repeated the results, and find with a yield of 145 p.ct. that the product contains 11.8 p.ct. N.

The reaction

$C_{12}H_{18}O_9 + 4HNO_3 - 4H_2O$

gives a tetranitrate with 11.5 p.ct. N and a yield of 159 p.ct. The ester reaction, therefore, is not simple. There are two sources of the loss of weight. The first of these is evident from the occurrence of certain secondary reactions which result in the solution of a certain proportion of the fibre substance in the acid mixture. To determine this quantitatively we have devised a suitable variation of the method of combustion with chromic acid (1st ed.).

The variation is required to meet the difficulty occasioned by the tension of the nitric acid and products of deoxidation. The mixed acids (10 c.c.), containing the organic by-products [Pg 132] in solution, are carefully diluted in a small flask with an equal volume of water, preventing rise of temperature. Nitrous fumes are evolved during the dilution. Strong sulphuric acid (15 c.c.) is now added, and the residue of nitrous fumes expelled by a current of air, agitating the contents of the flask from time to time. The combustion with CrO_3 is then proceeded with in the ordinary way. The gases evolved are measured (total volume) and calculated to C present in the form of products derived from the lignocellulose; and, assuming that this contains 47 p.ct. C, we may express the result approximately in terms of the fibre substance. The method was controlled by blank experiments, in which citric acid was taken as a convenient carbon compound for combustion. The C found was 34.9 p.ct. as against 34.3 p.ct. calculated. By this method we find that with maximum yields of nitrate at 143-145 p.ct. the organic matter in solution in the acid mixture amounted to 4.9 to 5.3 p.ct. of the original lignocellulose.

Introducing this quantity as a correction of the yield of nitrate in the original reaction, we must express the 143 parts as obtained from 95 of fibre substance instead of 100.

The yield per molecule $C_{12}H_{18}O_9$ (= 306) is therefore 462, whereas for a tetranitrate formed by a simple ester reaction the yield should be 486. The difference (24) represents 1.5 mol. H_2O split off by internal condensation.

The correction for total N is relatively small, raising it from 11.5 to 12.2, which remains in close agreement with the experimental numbers.

Monobenzoate. — Treated with the acid mixture yields a mixed nitrate. The yield is 130 p.ct., and the product contains 7.6 p.ct. $O.NO_2$ nitrogen. These numbers approximate to those required for reaction with $4HNO_3$ groups, three of the residues entering the cellulose, and one [Pg 133] (as NO_2) the benzene ring of the substituting group. For such a reaction the calculated numbers are: Yield 144 p.ct.; $O.NO_2$ nitrogen 7.1 p.ct.

The experimental numbers require correcting for the amount of loss in the form of products soluble in the acid mixture, viz. 7.6 p.ct.; but they remain within the range of the experimental errors sufficiently to show that the benzoyl group limits the number of OH groups taking part in the ester reaction to three. The corrected yield per 1 mol. of jute benzoate (410) is 576, as against the calculated 590 for $4HNO_3$ reacting. A loss of $1H_2O$ per molecule by internal condensation is therefore indicated.

Denitration. — The removal of the nitric groups from the esters is effected by digestion with ammonium sulphide. But the reactions are by no means simple. There is considerable hydrolysis of the lignocellulose to soluble products. Thus the *tetranitrate* yields only 46.4 of denitrated fibre in place of the calculated 66. The product is a cellulose, yielding only 0.5 per cent. furfural. The hydrolysed by-products, moreover, when freed from sulphur and distilled from hydrochloric acid, yielded only an additional 2.5 p.ct. furfural, calculated to the original lignocellulose.

These statistics confirm the evidence that the ester reaction is not simple. Such changes take place in the lignone-β-cellulose complex that they revert, not to their original form, but to soluble derivatives of different constitution. The mixed nitrate from the benzoate is denitrated to a cellulose amidobenzoate, which confirms the localisation of a nitro-group in the benzoyl residue.

(p. 157) **General Characteristics of the Lignocelluloses.** — Later investigations have somewhat modified and simplified our views of the constitution of the typical lignocellulose (jute), so far as this can be dealt with by the [Pg 134] statistics of its more important decompositions (original, pp. 157-161).

Cellulose. — There is little doubt that the furfural-yielding groups of the original are isolated in the form of the β-cellulose. Tollens emphasises this fact in his studies of cellulose-estimation methods. We had previously shown (original, p. 159) that the yield of furfural is not affected by the *chlorination*, but it appears from our numbers that only 50 p.ct. of these groups remain in the isolated cellulose, the residue undergoing hydrolysis to soluble compounds. In a carefully regulated hydrolysis following the chlorination it appears that the furfuroids are almost entirely conserved in the form of a cellulose.

Moreover, an investigation of the products dissolved by sodium sulphite solution from the chlorinated fibre has shown that they are practically free from furfuroids. This enables us to exclude the furfural-yielding groups from the lignone complex. At the same time, through our later studies of the hydroxyfurfurals, it is certain that these products are represented in the fibre substance and probably in the lignone complex.

Chlorination Statistics.—It has been pointed out by a correspondent—to whom we express our indebtedness—that we have made a mistake in calculating the proportion of lignone from the ratio of the Cl combining with the fibre substance or lignocellulose (p.ct), to that of the Cl *present in* the isolated lignone chloride (p.ct.). The lignocellulose combines with chlorine in the ratio 100: 8, but the lignone chloride *containing* 26.7 of chlorine means that, neglecting the hydrogen substituted, 73 of lignone combine with the 27 of chlorine approximately. On the uniform percentage basis the calculated proportion of lignone would be 8/37, or a little over 20 p.ct. [Pg 135]

In regard to the proportion of hydration attending the resolution, we have shown on constitutional grounds that this must be relatively small. Assuming approximately the formula $C_{19}H_{22}O_9$ for the lignone residue as it exists in combination, and the anhydride formula for the cellulose, these revised statistics now appear, as regards the carbon contents of the lignocellulose:

Cellulose, 44.4 C; lignone, 57.8.

$$80 \times 44.4 \div 100 = 35.52$$
$$20 \times 57.8 \div 100 = 11.56$$

47.08 p.ct. C in lignocellulose.

These conclusions are in accordance with the experimental facts, and, taken together with the new evidence we have accumulated from a study of the lignocellulose esters, we may sum up the constitutional points as follows: The lignocellulose is a complex of

Cellulose α	Cellulose β	Lignone
65 p.ct.	15 p.ct.	20
Allied to the normal celluloses	Yielding furfural approximately 50 p.ct.	One-third of which is of benzenoid type

The lignone contains but little hydroxyl. The celluloses are in condensed hydroxyl union with the lignone, but the combination occurs by complexes of relatively large molecular weight.

DIE CHEMIE DER LIGNOCELLULOSEN—EIN NEUER TYPUS.

W. C. Hancock and O. W. Dahl (Berl. Ber., 1895, 1558).

Chemistry of Lignocelluloses—A New Type.

The stem of the aquatic *Æschynomene aspera* offers an exceptional instance of structural modification to serve the special function of a 'float,' 1 grm. of substance occupying an apparent volume of 40-50 c.c. This pith-like substance is [Pg 136] morphologically a true wood (De Bary), and the author's investigations now establish that it is in all fundamental points of chemical composition a lignocellulose, although from its colour reactions it has been considered by botanists to be a cellulose tissue containing a proportion of lignified cells. Thus the main tissue is stained blue by iodine in presence of hydriodic acid (1.5 s.g.), and the colour is not changed on washing. The ordinary lignocelluloses are stained a purple brown changed to brown on washing. The reactions with phloroglucol and with aniline salts, characteristic of these compounds, is only faintly marked in the main tissue, though strongly in certain individual cells.

The following quantitative determinations, however, establish the close similarity of the product to the typical lignocelluloses:

Elementary Analysis.—C 46.55, H 6.7. *Furfural* 11.6 p.ct., of which there remained in the residue from alkaline hydrolysis (71 p.ct.) 8.0, i.e. about 70 p.ct. The distribution of the furfuroids is therefore not affected by the alkaline treatment.

Chlorination.—The substance (after alkaline hydrolysis) takes up 16.9 p.ct. Cl, of which approximately one-half is converted into hydrochloric acid.

Methoxyl.—O.CH$_3$ estimated = 2.9 p.ct.

Ferric Ferricyanide Reaction.—Increase of weight due to blue cyanide fixed (1) 75 p.ct., (2) 96 p.ct. Ratio, Fe: CN = 1: 2, 4.

Hydroxyl Reactions.—In the formation of nitric esters and in the sulphocarbonate reaction the substance gave results similar to those obtaining for the jute fibre.

These results establish the general identity of this peculiar product of plant life with the lignocelluloses, at the same time that they show that certain of the colour reactions supposed to characterise the lignocelluloses are due to by-products which may or may not be present. [Pg 137]

(p. 172) **Composition of Elder Pith.**—In a systematic investigation of the celluloses in relation to function we shall have to give special attention to the parenchymatous tissues of all kinds. These are, for structural reasons, not easily isolated, for which reason and their generally 'inferior' functions they do not present themselves to chemical observation in the same obvious way as do their fibrous relatives. The pith of the elder, however, *is* readily obtained in convenient masses, and a preliminary investigation of the entire tissue has established the following points:

The *reactions* of the tissue are in all respects those of the lignocelluloses.

Composition.—Ash, 2.2 p.ct.; moisture in air-dry state, 12.3 p.ct. Alkaline hydrolysis (loss): (*a*) 14.77, (*b*) 17.84. Cellulose (yield), 52.33 p.ct. Nitrate-reaction complicated by secondary reactions and yields low, 90.95 p.ct. *Sulphocarbonate reaction:* Resists the treatment, less than 10 p.ct. passes into solution.

Furfural. — The original tissue yields 7.13 p.ct.; the residue from alkaline hydrolysis (*b*) 5.40 p.ct.

This tissue is, therefore, a lignocellulose having the chemical characteristics typical of the group, but of less resistance to hydrolytic actions.

The investigation will be prosecuted in reference to the cause of differentiation in this latter respect. Probably the pectocelluloses are represented in the tissue.

The Insoluble Carbohydrates of Wheat (grain).

H. C. Sherman (J. Amer. Chem. Soc., 1897, 291).

(p. 171) This is a study of the constituents of the cell-walls of wheat grain. Bran was taken as the most convenient form of the raw material, being freed from starch by treatment with malt extract, and further treated (1) with cold dilute ammonia, [Pg 138] (2) cold dilute soda lye (2 p.ct. NaOH), and (3) boiling 0.1 p.ct. NaOH. The product retained only 1.25 p.ct. proteids, and yielded 15.62 p.ct. furfural.

Acid Hydrolysis. — The product was boiled 30 mins. with dilute acid (1.25 p.ct. H_2SO_4), and the solution boiled until the Fehling test showed no further increase of monoses. At the limit the reducing power of the dissolved carbohydrates was 91.3 p.ct., that of dextrose. Converted into osazones the analysis showed them to be *pure pentosazones*. The *hemicellulose* of wheat is, therefore, according to the author, *pure pentosane*.

Residue. — This was a lignocellulose yielding 11.5 p.ct. furfural. It was subjected to a series of treatments with ferric ferricyanide, and the proportion of Prussian blue fixed was determined by increase of weight, viz. from 10 p.ct. to 47 p.ct. according to the conditions. The results confirmed those of Cross and Bevan first obtained with the typical lignocellulose (jute).

Chlorination.-The residue was boiled with dilute alkali, washed, and exposed to chlorine gas. The resulting lignone chloride was isolated by solution in alcohol, &c. It yielded 26.7 p.ct. Cl on analysis. In this and its properties it appeared to be identical with the product isolated by Cross and Bevan from jute, with the empirical formula $C_{19}H_{18}Cl_4O_9$.

Cellulose was isolated from the residue by three of the well-known methods, and the following comparative numbers are noteworthy:

Method	F. Schulze Dil. HNO_3 $KClO_3$	Lange Fusion KOH	Cross and Bevan Chlorine, &c.
Cellulose p.ct. obtained	66.0	39.3-43.1	66.5
Furfural p.ct. of cellulose	7.0	3.96	5.62
Residual nitrogen	0.22	0.03	0.00
Ferricyanide reaction, Prussian blue fixed	6.04	0.89	0.92

[Pg 139]

The author remarks: 'It is evident no one feature can be urged as a criterion in judging between the methods, but all must be taken into consideration. Such a comparison shows the superiority of the chlorination method.'

The cellulose is not of the normal (cotton) type, since on treatment with sulphuric acid it dissolves with considerable discolouration, but only to the extent of about 80 per cent. The dissolved monoses converted into osazones were found to consist of hexoses only. The cellulose treated with caustic soda solution (5 p.ct. NaOH) in the cold yielded 20 p.ct. of its weight of soluble constituents, but as the residue yielded 3.34 p.ct. furfural the attack of the alkali is by no means confined to the furfuroids.

Animal Digestion of the Constituents of Bran.—Observations on a steer fed upon wheat bran only established the following percentage digestion of the several constituents:

Soluble carbohydrates	96.9
Starch	100.0
Free pentosanes	60.2
Cellulose	24.8
Lignin complex	36.7
Proteid	82.96
Ether extract	42.73
Nitrogen-free extract	76.08
Crude fibre	32.21

JOURNAL OF THE IMPERIAL INSTITUTE

(Research Department, Vols. 1-2, 1895-6).

(p. 109) In this journal appear a series of notices of the results of analyses of vegetable fibres by the method described in 'Report on Miscellaneous Fibres' (Col. Ind. Exhibition [Pg 140] Reports, p. 368) [C. F. Cross]. These investigations deal with the following subjects:

1895. p. 29 Various Indian Fibres—more particularly Sida.
(a) 118 (a) Fibres from Victoria; (b) SpecialAnalyses ofSamples of Jute; (c) Paper-making Fibres from S. Australia.
202   Fibres from Victoria.
287   Fibres from Victoria.
366   Sisal from Trinidad.
373   Rope-fibres from Grenada.
(b) 398 Report of Experiments on Indian Jute (1).
435}   Fifth and Sixth Report on Australian Fibres.
473}
1896. 68 Hibiscus and Abroma Fibres.
104-5   Hibiscus, Urena, and Crotalaria Fibres.
141   Indian Sisal

(c) 182-3 Report of Experiments on Indian Jute (2).
264   Sanseviera from Assam.

From the above we may draw the general conclusion that the scheme of investigation has been found in practice to answer its main purpose, viz. to afford such numerical constants as determine industrial values. In illustration we may cite (a) the results of analyses of specially selected samples of jute, from which it will be seen that there is a close concordance of value as ordinarily determined from external appearance, with the chemical constants as determined in the laboratory.

	Quality of Jute			
	Low	Medium	Extra	Extra Fine
Moisture	11.0	10.4	11.1	9.6
Ash	0.87	2.8	1.0	0.7
Alkaline hydrolysis (a) 5 mins. boiling	13.2	11.6	8.5	9.1
Alkaline hydrolysis (b) 60 mins. boiling	16.1	17.5	12.5	13.1
Mercerising treatment	9.2	10.5	10.3	8.5
Nitration (increase p.ct.)	36.6	35.7	37.5	36.7
Cellulose (yield)	71.4	70.0	79.0	77.7
Acid purification	2.6	1.3	1.9	2.0

[Pg 141]

A useful series of experiments, initiated by the Institute, is that noted under (b) and (c) above.

(1) To ascertain the quality of the fibre extracted from the plant at different stages of growth, quantities of 400 lbs. of the stalks were cut at successive stages and the fibre isolated after steeping 14-20

days. The fibre was shipped to England and chemically investigated, with the following results:

No. 1. Cut before appearance of inflorescence.
"   2. " after budding.
"   3. " in flower.
"   4. " after appearance of seed-pod.
"   5. " when fully matured.

	(1)	(2)	(3)	(4)	(5)
Moisture	11.55	8.74	10.7	10.0	9.72
Ash	1.1	1.1	1.1	1.1	0.90
Alkaline hydrolysis (a)	6.2	8.5	9.7	8.9	7.3
" " (b)	10.5	11.9	11.6	12.0	11.2
Mercerising treatment	10.2	10.7	12.0	8.1	11.0
Nitration	37.2	32.1	32.2	33.2	36.6
Cellulose	74.0	76.2	74.1	74.8	76.4
Acid purification	0.8	0.5	0.7	2.4	1.4

It will be thus seen that there are no changes of any essential kind in the chemical composition of the bast fibre throughout the life-history of the plant, confirming the conclusion that the 'incrustation' view of lignification is consistent only with the structural features of the changes, and so far as it has assumed the gradual overlaying of a cellulose fibre with the lignone substance it is not in accordance with the facts.

Examination of the samples from the point of view of textile quality showed a superiority of No. 1 in fineness, softness, and strength; from this stage there is observed a progressive deterioration, but the No. 4 sample (which was taken at the usual period of cutting) is superior to No. 5. [Pg 142]

In a further series of experiments (c) the jute was subjected to certain chemical treatments immediately after the separation of the fibre from the plant. These consisted in steeping (1) in solution of sodium carbonate, as well as of plant ashes, and (2) in sulphite of soda, the purpose of the treatments being to modify or arrest the changes which take place in the fibre when press-packed in bales for shipment. The samples were shipped from India under the usual conditions and examined soon after arrival. It was found that the chemical treatments had produced but small changes in chemical composition of the fibre-substance. The sulphite treatment was the more marked in influence, somewhat lowering the cellulose and nitration constants. The conclusion drawn from the results was that they afford no prospect of any useful modification, i.e. improvement of the textile quality of the fibre by any chemical treatments such as could be applied to the fibre on the spot before drying for press-packing and shipment.

The other matters investigated in the Institute laboratory and reported on as indicated above are rather of commercial significance, and contributed no points of moment to the chemistry of cellulose.

OBSERVATIONS ON SOME OF THE CHEMICAL SUBSTANCES IN THE TRUNKS OF TREES.

F. H. Storer (Bull. Bussey Inst., 1897, 386).

(p. 172) An examination of the outer and inner wood and of the bark of the grey birch, at different seasons of the year, gave the following yields of furfural p.ct. on the dry substance: [Pg 143]

	Wood		Bark
	Inner	Outer	
May	21.3	19.6	16.7
July	16.6	18.8	11.4
October	16.2	16.3	12.3

The paper contains the results of treating the woods and various vegetable products with hydrolysing agents in order of intensity: (*a*)

Malt-extract at 60°C., (*b*) boiling dilute HCl (1.0 p.ct. HCl), and (*c*) boiling dilute HCl (2.5 p.ct.). The residues were found to yield considerable proportions of furfural. The following numbers are typical:

	Birch		Stones of		
	Bark	Wood	Date	Apricot	Peach
Action of malt extract calculated as starch dissolved	4.24	3.5	5.2	1.5	—
			Mannan		
Residue boiled, 1 p.ct. HCl gave pentosanes dissolved.	—	—	11.7	14.1	6.7
Residue yielded furfural	19.3	17.8	3.4	9.6	9.7

The proportion of pentosanes (furfuroids) removed, i.e. hydrolysed by boiling with hydrochloric acid of 2.5 p.ct. HCl, is shown by the following estimations of furfural:

	Birch		Sugar maple		Apricot stones
	Bark	Wood	Outer wood	Inner wood	
In original substance	16.7	19.6	18.2	20.7	18.4
In residue from action of 2.5 p.ct. HCl	6.53	8.6	4.9	6.4	7.0

Wood Gum.—The paper contains some observations on the various methods of isolating this product. Attention is directed to the necessary impurity of the product, and to the fact that the numbers for furfural and for the xylose yielded by hydrolysis are considerably less than for a pure pentosane. [Pg 144]

Estimation of Cellulose.—The author investigated the process of Lange and the 'celluloses' obtained from various raw materials. The products from the woods of birch and maple contained furfural-yielding constituents, represented by yields of 6-8 p.ct. furfural. Preference is given to the process by comparison with others, at the same time that it is recommended in all cases to examine the product for furfural quantitatively, converting the numbers into pentosane equivalents, and subtracting from the total 'cellulose' to give the true cellulose.

ZUR KENNTNISS DER MUTTERSUBSTANZEN DES HOLZ-GUMMI.

E. Winterstein (Ztschr. Physiol. Chem., 1892, 381).

ON THE MOTHER SUBSTANCES OF WOOD-GUM.

(p. 188) According to the text-books beech-wood may be regarded as the typical raw material for the preparation of the laboratory product known as wood-gum. The author has subjected beech-wood and beech-wood cellulose (Schulze process) to a range of hydrolytic treatments, acid and alkaline, in order to determine the conditions of selective action upon the mother substance of the wood-gum. In the main it appears that this group of furfuroids is equally resistant with the cellulose constituents of the wood; in fact, that the mother substance of wood-gum is a modified cellulose, and exists in the wood in chemical combination with the 'incrusting substances.'

Of the author's experimental results the following may be cited as typical:

Substance	Yield of furfural p.ct.
Original beech-wood	13.8
After boiling 3 hrs. with 1.25 p.ct. H_2SO_4 (residue)	10.1
" " " " 5.0 " " "	5.6

[Pg 145]

Cellulose	— isolated by Schulze process (yield 53 p.ct.)	6.9
"	after further 14 days' digestion with the Schulze acid (HNO_3 + $KClO_3$)	5.9
"	after extraction with 5 p.ct. NaOH in cold (residue)	5.0
"	after second extraction with 5 p.ct. NaOH in cold (residue)	4.4

UEBER DIE FRAGE NACH DEM URSPRUNG UNGESÄTTIGER VERBINDUNGEN IN DER PFLANZE.

C. F. Cross, E. J. Bevan, and C. Smith (Berl. Ber., 1895, 1940).

ON THE SOURCE OF THE UNSATURATED COMPOUNDS OF THE PLANT.

(p. 179) In distilling for furfural by the usual methods of boiling cellulosic products with condensing acids, the furfural is accompanied by volatile acids, also products of decomposition of the cellulosic complex. A series of distillations was carried out with dilute sulphuric acids of varying concentration from 10-50 H_2SO_4: 90-50 H_2O by weight, using barley straw as a typical cellulosic material. The distillates were collected in successive fractions, and the furfural and volatile acid determined. The results are given in the form of curves. The aggregate yields were as follows: —

Concentration of acid (H_2SO_4) p.ct.	10	15	20	30	40	50
Furfural yield p.ct. of straw	2.0	2.0	4.4	10.1	11.5	11.0
Volatile acid (calculated as acetic acid) p.ct. of straw	1.7	1.9	3.1	4.3	6.3	14.8

With acids up to 20 p.ct. H_2SO_4 both products are formed concurrently and in nearly equal quantity. With the 30 p.ct. [Pg 146] acid there is a great increase in the total furfural, and with the 40 p.ct.

acid it reaches nearly the maximum obtainable with HCl of 1.06 s.g. (Tollens), in this case 12.4 p.ct. The volatile acid increases, but in less ratio; it is also produced concurrently. With 50 p.ct. H_2SO_4 the conditions are changed. The total furfural is rapidly formed, whereas the volatile acid continues to be formed long after the aldehyde ceases to come over. Moreover, whereas in the previous cases it was mainly acetic acid, it is now mainly formic acid. The method was then extended to a typical series of celluloses, heated with the more concentrated acid (40-50 p.ct. H_2SO_4), with the following results:

—	—	Volatile acid	
		Acetic	Formic
Swedish filter-paper	0.3	2.7	17.2
Esparto cellulose	12.4	3.2	16.6
Bleached cotton	trace	3.1	13.2
Raw cotton (American)	—	5.0	9.4
Jute cellulose	5.2	4.9	22.7
Beech (wood) cellulose	6.4	3.5	14.6

The tendency in the hexoses and their polyanhydrides to split off one carbon atom in the oxidised form, throws some light on the furfurane type of condensation, which is represented in the lignocelluloses. We are still without any evidence as to the possible transition of the hexoses to benzenoid compounds. Such transitions would be more easily explained on the assumption that the celluloses are composed in part of polyanhydrides of the ketoses.

SPIRITUS AUS CELLULOSE UND HOLZ.

E. Simonsen (Ztschr. angew. Chem., 1898, 3).

PRODUCTION OF ALCOHOL FROM CELLULOSE AND WOOD.

(pp. 50, 209) This investigation was undertaken with one main object—to determine the optimum conditions of treatment [Pg 147] of wood-cellulose and of wood itself for conversion into 'fermentable sugar.' The process of 'inversion' or hydrolysis, by digestion with dilute acid at high temperature, involves the four main factors: pressure (i.e. temperature), concentration of acid, ratio of liquid to cellulose and duration of digestion. Each of these was varied in definite gradations, and the effect measured. The degree of action was measured in terms of 'reducing sugar,' calculated from the results of estimation by Fehling solution, as 'glucose' per cent. of original cellulose (or wood).

(*a*) *Cellulose*. [Wood-cellulose obtained by bisulphite process.]—With a proportion of total liquid to cellulose of 27:1, and using sulphuric acid as the hydrolysing agent, the optimum results were obtained with acids of 0.45-0.60 p.ct. (H_2SO_4) and pressures of 6-8 atm. The maximum yield of 'sugar' was 45 p.ct. of the cellulose.

Under the above conditions the maximum of conversion is attained in 2 hours.

Having now regard to the production of a solution of maximum *concentration* of dissolved solids, the following conditions were asertained to fulfil the requirement, and, in fact, may be regarded as the economic optimum:

Proportion of total liquid	6 times wt. of cellulose
Concentration of acid	0.5 p.ct. H_2SO_4
Pressure	10 atm.
Duration of digestion	1.5 hour

giving a yield of 41 p.ct. 'reducing sugar' calculated to the original cellulose (dry).

Alcoholic Fermentation of Neutralised Extract.—The liquors were found to ferment freely, and on distillation to yield a quantity of alcohol equal to 70 p.ct. of the theoretical—i.e. on the basis of the numbers for copper oxide reduction.

(b) *Hydrolytic 'Conversion' of Wood (Lignocellulose).* — A [Pg 148] similarly systematic investigation carried out upon pine sawdust established the following as optimum conditions:

Proportion of total liquid	5 times wt. of wood
Concentration of acid	0.5 p.ct. H_2SO_4
Pressure	9 atm.
Duration of digestion	15 minutes

giving a yield of 20 p.ct. 'reducing sugar,' calculated from the 'Fehling' test.

Fermentation of the neutralised extracts gave variable results. The highest yields obtained were 60 p.ct. of theoretical, the author finally concluding that under properly controlled conditions of inversion and fermentation 100 kg. wood yield 6.5 l. absolute alcohol.

ÜBER DIE URSACHE DER VON SIMONSEN BEOBACHTETEN UNVOLLSTÄNDIGKEIT DER VERGÄHRUNG DER AUS HOLZ BEREITETEN ZUCKERFLÜSSIGKEITEN.

B. Tollens (Ztschr. angew. Chem., 1898, 15).

ON THE CAUSE OF INCOMPLETE FERMENTATION OF SUGARS OBTAINED BY ACID HYDROLYSIS OF WOOD.

The author criticises Simonsen's explanation of the results obtained with extracts from pine wood. The incompleteness of fermentation of the products is certainly due in part to the presence of furfural-yielding carbohydrates, which are resistant to yeast. The pine woods contain 8-10 p.ct. of these constituents in their anhydride form ('pentosanes'). They yield readily to acid hydrolysis, and certainly constitute a considerable percentage of the dissolved products. A similar complex was obtained by the author in his investigation of peat (Berl. Ber. 30, 2571), and was found to be similarly incompletely attacked by yeast. The yields of alcohol corresponded [Pg 149] with the proportion of the total carbohydrates disappearing. These were the hexose constituents of the hydrolysed complex, the pentoses (or 'furfuroids') surviving intact.

UEBER SULFITCELLULOSEABLAUGE.

H. Seidel (Ztschr. angew. Chem., 1900).

WASTE LIQUORS FROM BISULPHITE PROCESS.

(p. 210) Later researches confirm the conclusion that in the soluble by-products of these cellulose processes the S is combined as a SO_3H group. The following analyses of the isolated lignin sulphonic acid are cited:

	C	H	S
(a) Lindsey and Tollens	56.12	5.30	5.65
(b) Seidel (1)	56.27	5.87	5.52
(c) Seidel and Hanak (2)	53.69	5.22	8.80
(d) Street	50.22	5.64	7.67

The variations are due to the varying conditions of the digestion of the wood and to corresponding degrees of sulphonation of the original lignone group. Calculating the composition of the latter from the above numbers on the assumption that the S represents SO_3H, the following figures result:

	(a) and (b)	(c)	(d)
C	64.00	65.1	59.61
H	6.65	6.33	6.69

This author considers that beyond the empirical facts established by the above named [10] very little is yet known in regard to the constitution of the lignone complex. [Pg 150]

Nor is there any satisfactory application of this by-product as yet evolved. Evaporation and combustion involve large losses of sulphur [D.R.P. 74,030, 83,438; Seidel and Hanak, Mitt. Techn. Gew. Mus. 1898]. A more complete regeneration of the sulphur has been the subject of a series of patents [D.R.P. 40,308, 69,892, 71,942, 78,306, 81,338], but the processes are inefficient through neglect of the actual state of combination of the S, viz. as an organic sulphonate. The process of V.B. Drewson (D.R.P. 67,889) consists in heating with lime under pressure, yielding calcium monosulphite (with sulphate and the lignone complex in insoluble form). The sulphite is redissolved as bisulphite by treatment with sulphurous acid. This process is relatively costly, and yields necessarily an impure lye. It has been proposed to employ the product as a foodstuff both in its original form and in the form of benzoate (D.R.P. 97,935); but its unsuitability is obvious from its composition. A method of destructive distillation has been patented (D.R.P. 45,951). The author has investigated the process, and finds that the yield of useful products is much too low for its economical development. Fusion with alkaline hydrates for the production of oxalic acid (D.R.P. 52,491) is also excluded by the low yield of the product.

The application of the liquor for tanning purposes (D.R.P. 72,161) appears promising from the fact that 28 p.ct. of the dry residue is removed by digestion with hide powder. This application has been extensively investigated, but without practical success. Various probable uses are suggested by the viscosity of the evaporated extract. As a substitute for glue in joinery work, bookbinding, &c., it has proved of little value. It is applied to some extent as a binding material in the [Pg 151] manufacture of briquettes, also as a substitute for gelatin in the petroleum industry. Cross and Bevan (E.P. 1548/1883) and Mitscherlich (D.R.P. 93,944 and 93,945) precipitate a compound of the lignone complex and gelatin by adding a solution of the latter to the liquors. The compound is redissolved in weak alkaline solutions and employed in this form for engine-sizing papers. Ekman has patented a process (D.R.P. 81,643) for 'salting out' the lignone sulphonates, the product being resoluble in water and the solution having some of the properties of a solution of dextrin. Owing to its active chemical properties this product—'dextron'—has a limited capability of substituting dextrin. The suggestion to employ the evaporated extract as a reducing agent in indigo dyeing and printing has also proved unfruitful. The author's application of the soda salt of the lignone sulphonic acid as a reducing agent in chrome-mordanting wool and woollen goods (D.R.P. 99,682) is more successful in practice, and its industrial development shows satisfactory progress. The product is known as 'lignorosin.'

FOOTNOTES:

[10] See more particularly: Lindsey and Tollens, *Annalen*, 267, 341; Cross and Bevan's *Cellulose*, pp. 197-203; Street, Inaug.-Diss., Göttingen, 1892; Klason, *Rep. d. Chem. Ztg.* 1897, 261; Seidel and Hanak, *Mitt. d. Techn. Gew. Mus.* 1897-1898.

[Pg 152]

SECTION VII. PECTIC GROUP

UNTERSUCHUNGEN ÜBER PECTINSTOFFE.

R. W. Tromp de Haas and B. Tollens (Lieb. Ann., 286, 278).

ÜBER DIE CONSTITUTION DER PECTINSTOFFE, B. Tollens (*ibid.* 292).

INVESTIGATIONS OF PECTINS.

(p. 216) It is generally held that the pectins are, or contain, oxidised derivatives of the carbohydrates. The authors have isolated and analysed a series of these products, and the results fail to confirm a high ratio O : H. The following are the analytical numbers:

Pectin from	Ash	C	H	Ratio H : O
Apple	6.2	43.4	6.4	1 : 7.9
Cherry	20.5	42.5	6.5	1 : 7.9
Rhubarb	4.2	43.3	6.8	1 : 7.4
Currant	5.0	47.1	5.9	1 : 8.5
Greengage	3.3	43.0	5.9	1 : 8.5
Turnip	7.3	41.0	5.9	1 : 9.0

Acid hydrolysis (4 p.ct. H_2SO_4) gave syrupy products not crystallisable — in certain cases the hydrolysis was accompanied by separation of insoluble cellulose. The insoluble product from currant pectin had the composition C 54.4, H 5.0.

Tollens points out that the results of empirical analysis are inconclusive; and that from the acid reactions of these products and their combination with bases, carboxylic groups are present, though probably in anhydride or ester form.

The pectins may be regarded as closely related to the mucilages (*Pflanzenschleim*), differing from them only by the presence of the oxidised groups in question.

UEBER DIE CONSTITUTION DER PECTINSTOFFE.

C. F. Cross (Berl. Ber., 1895, 2609).

CONSTITUTION OF PECTINS.

It is pointed out that the composition of the pectin of white currants, as given in the preceding paper, is that of the typical lignocellulose, the jute fibre. The product was isolated and further investigated by the author. It gave 9.8 p.ct. furfural on boiling with HCl (1.06 s.g.), reacted freely with chlorine, giving quinone chlorides, and with ferric [Pg 153] ferricyanide to form Prussian blue. This 'pectin' is therefore a form of soluble lignocellulose. The 'pectic' group consequently must be extended to include hydrated and soluble forms of the mixed complex of condensed and unsaturated groups with normal carbohydrates, such as constitute the fibrous lignocelluloses.

UEBER DAS PFLANZLICHE AMYLOID.

E. Winterstein (Ztschr. Physiol. Chem., 1892, 353).

ON VEGETABLE AMYLOID.

(p. 224) A group of constituents of many seeds, distinguished by giving slimy or ropy 'solutions' under the action of boiling water are designated 'amyloid.' They are reserve materials, and in this, as in the physical properties of their 'solutions,' they are very similar to starch. They are, however, not affected by diastase; and generally are more resistant to hydrolysis. Typical amyloids have been isolated by the author from seeds of *Tropæolum majus, Pæonia officinalis*, and *Impatiens Balsamina*. The raw material was carefully purified by exhaustive treatment with ether and alcohol, &c.; the amyloid then extracted by boiling with water, and isolated by precipitation with alcohol. Elementary analysis gave the numbers C 43.2, H 6.1. On boiling with 12 p.ct. HCl it gave 15.3 p.ct. furfural; oxidised with nitric acid it yielded 10.4 p.ct. mucic acid. Specimens from the two first-named raw materials gave almost identical numbers.

Hydrolysis.—On boiling with dilute acids these products are gradually broken down, dissolving without residue. In this respect they are differentiated from the mucilages, which give a residue of cellulose (insoluble). From the solution the author isolated crystalline galactose, but failed to isolate a pentose. Dextrose was also not identified directly.

[Pg 154]

The tissue residues left after extracting the amyloid constituent, as above described, were subjected to acid hydrolysis. A complex of products was obtained, from which galactose was isolated. A furfural-yielding carbohydrate was also present in some quantity, but could not be isolated. The original seed tissues, therefore, contain an amyloid and a hemicellulose, the latter differentiated in its resistance to water. Both yield, however, to acid hydrolysis a complex of products of similar composition and constitution.

UEBER DEN GEHALT DES TORFES AN PENTOSANEN ODER FURFUROLGEBENDEN STOFFEN UND AN ANDEREN KOHLENHYDRATEN.

H. v. Feilitzen and B. Tollens (Berl. Ber., 1897, 2,571).

CARBOHYDRATE CONSTITUENTS OF PEAT.

(p. 240) An investigation of typical peats taken at successive depths showed increasing percentage of carbon, and inversely a decreasing yield of furfural. The numbers may be compared with those for *Sphagnum cuspidatum*—with C = 49.80 p.ct., and furfural 7.99 p.ct., calculated to dry, ash-free substance:

Depth at which taken		C p.ct.	Furfural p.ct.
I.	20-100 cm.	51.08	6.93
	100-200 "	53.52	5.30
	200-300 "	58.66	3.19
II.	Surface-20 "	55.47	3.40
	20-60 "	55.06	3.48
	60-100 "	58.25	1.45
	100-120 "	58.23	1.19
	180-200 "	57.57	1.80

Cellulose was estimated by the Lange method. The yield from *Sphagnum* was 21.1 p.ct.

From specimen I. at { 20-100 cm.   15.20

{100-200 "   6.87

[Pg 155]

From the peat of lower depths no cellulose could be obtained.

Hydrolysis (acid).—On heating with 1 p.ct. H_2SO_4 at 130-135°, soluble carbohydrates were obtained, amongst which mannose was identified, and galactose shown to be present in some quantity. After fermenting away the hexoses, the residue was treated with phenylhydrazine and an osazone separated. It contained 17.3 p.ct. N, but melted at 130°. The substance could not be identified as an osazone of any of the yet known pentoses.

SECTION VIII. INDUSTRIAL AND TECHNICAL. GENERAL REVIEW

The Industrial Uses of Cellulose.

C. F. Cross (Cantor Lectures, Soc. of Arts, 1897).

(p. 273) A series of three lectures, in which the more important industries in cellulose and its derivatives are dealt with on their scientific foundations, and by means of a selection of typical problems. In reference to textiles, the small number of vegetable fibres actually available, out of the endless variety afforded by the plant world, is referred to the number of conditions required to be fulfilled by the individual fibre, thus: yield per cent. of harvested weight or per unit of field area, ease of extraction, the absolute dimensions of the spinning unit, and the proportion of variation from the mean dimensions; the relative facility with which the unit fibre can be isolated preparatory to the final twisting operation; the chemical constants of the fibre substance, especially the percentage of cellulose and degree of resistance to hydrolysis. It is suggested that any important addition to the very limited number fulfilling the conditions, or any great improvement in these, can only result from very elaborate artificial selection and cultural developments on this basis. [Pg 156]

The paper making fibres are shown to fall into a scheme of classification based on chemical constitution, and consisting of the four groups: (*a*) Cotton [flax, hemp, rhea], (*b*) wood celluloses, (*c*) esparto, straw, and (*d*) lignocelluloses. Papers being exposed to the natural disintegrating agencies, more especially oxygen, water (and hydrolysing agents generally), and micro-organisms, the relative resistance of the above groups of raw materials is discussed as an important condition of value. The indirect influence of the ordinary sizing and 'filling' materials is discussed. The paper-making quality of the fibrous raw materials is also discussed, not merely from the point of view of the form and dimensions of the ultimate fibres, but their capacity for 'colloidal hydration.' This is complementary to the action of rosin, i.e. resin acids, in the engine-sizing of papers; and the proof of the potency of this factor is seen in the superior effects obtained in sizing jointly with solutions of cellulose and, more particularly, viscose and rosin. Wurster's much-cited monograph of the subject of rosin-sizing ['Le Collage des Papiers,' Bull. Mulhouse, 1878] neglects to take into consideration the contribution of the cellulose hydrates to the total and complex sizing effect, and hence gives a partial view only of the function of the resin acids.

In further illustration of fundamental principles various developments in the textile industries are discussed, e.g. the bleaching of jute, cotton, and flax, and special developments in the spinning of rhea and flax.

The concluding lecture deals with later progress in the industrial applications of cellulose derivatives, chiefly the sulphocarbonate (viscose); the nitrates, in their applications to explosives, on the one hand, and the spinning of artificial fibres (lustra-cellulose), on the other; and the cellulose acetates. [Pg 157]

La Viscose et le Viscoide.

C. H. Bardy (Bull. Soc. d'Enc. Ind. Nationale, 1900, March).

This is a report presented to the Committee of Economic Arts of the above Society, dealing with the industrial progress in products obtained by means of the sulphocarbonate of cellulose (viscose).

The following developments are noted:

Engine-sized Papers.—The viscose, by coating the fibres with regenerated cellulose hydrate, adds very much to the tensile strength of papers. Increase of 40-60 p.ct. is attainable by addition of cellulose in this form from 1-4 p.ct. on the weight of the paper.

Viscoid.—Solid aggregates are formed by incorporating viscose with mineral matters, hydrocarbons, &c. Products are cast or moulded into convenient forms, and, after purification and sufficient ageing, are available for various structural uses.

Paint.—The viscose is used as a vehicle for pigments, the mixture being used either as a paint or for coating papers with fine surfaces, such as required in the reproduction of photo-blocks. In these applications the extraordinary viscosity of the product conditions the economic use of the cellulose in competition with oils, on the one hand, and organic colloids, such as gelatine, casein, &c., on the other.

By suitable alteration of the formula for making the paint a product is obtained which has an extraordinary power of removing paint from old painted surfaces. The product has been officially adopted by the French Admiralty, and receives extensive application in removing the paint from ships.

Films.—Films are produced from the viscose itself in various ways. Plane or flat by solidifying the viscose on glass surfaces, removing the by-products and rolling the films. The [Pg 158] film is also produced by applying the viscose on textile fabrics, drying down, and fixing on a stenter machine, then washing away the alkaline by-products from the fixed film. A large number of industrial effects are obtained by suitably varying the mixtures applied.

Cellulose-indiarubber.—The viscose, in its concentrated form, can be incorporated with rubber-hydrocarbon mixtures, and these mixtures can be used both as water-proofing films, as applied to textiles, or can be solidified into the class of goods known as 'mechanicals.' The cellulose not only cheapens the mixture, but produces new technical effects.

Spinning.—The viscose is spun by special methods, patented by C. H. Stearn. As produced in thread form, the diameters are approximately those of natural silk. In commercial form it is a multiple thread (of 15 or more units) at from 50-200 deniers on the silk counts. It is a thread of high lustre, and more nearly approaches the normal cellulose in chemical properties than any of the other artificial silks. It can also be spun in threads of very much larger diameter, which can be used as a substitute for horsehair, for carbonising for incandescent electric lamps, &c.

Cellulose Esters.—These are conveniently made from cellulose, regenerated from the solution as sulphocarbonate. The tetracetate is made from this product on the industrial scale. Nitrates are conveniently made by treatment with the ordinary mixed acids. For fuller details the original report may be consulted.

VISKOS.

R. W. Strehlenert (Svensk Kemisk Tidskrift, Stockholm, 1900, p. 185).

A report on the industrial development of viscose, covering essentially the same ground as the above. [Pg 159]

Ueber die Viscose.

B. M. Margosches (Reprint from Zeitschrift für die gesammte Textil-Industrie, 1900-01, Nos. 14-20). [11]

Report of Committee on the Deterioration of Paper.

(Soc. of Arts, 1898.)

(p. 304) The Report of a Representative Committee appointed by the Society of Arts to inquire into the question of qualities of book papers in relation to their several applications, and more especially for documents of permanent value.

The report first discusses the two directions of depreciation of papers in use: (1) Actual disintegration shown by loss of resistance to fracture by simple strain, and by loss of elasticity—i.e. increase of brittleness; (2) discolouration. These are independent effects, but often concurrent. They are the result of chemical changes of the cellulose basis of the paper, brought about by acids or oxidants used in the process of manufacture, and not completely removed from the pulp, or by acid products of bleaching—e.g. oxycelluloses or chlorinated derivatives; again, by the changes of starch used as a 'sizing' agent, or by oxidations induced by rosin constituents when the rosin is used in excess. Discolouration is an attendant phenomenon of these changes, but is more frequently due to the presence of the lower-grade celluloses (esparto and straw) and the lignocelluloses (mechanical wood-pulp).

The physical and chemical qualities of papers depending primarily upon their fibrous or pulp basis, and in a secondary degree upon the kind and proportion of the constituents added [Pg 160] for the purpose of filling and 'sizing,' the report concludes with the following recommendations, positive and negative, under these heads:

The Committee find that the practical evidence as to permanence fully confirms the classification given in the Cantor Lectures on 'Cellulose,' 1897 [J. Soc. Arts, xlv. 690-696], and which ranges the paper-making fibres in four classes:

(A) Cotton, flax, and hemp (rhea).

(B) Wood celluloses, (*a*) sulphite process and (*b*) soda and 'sulphate' process.

(C) Esparto and straw celluloses.

(D) Mechanical wood-pulp.

In regard, therefore, to papers for books and documents of permanent value, the selection must be taken in this order, and always with due regard to the fulfilment of the conditions of normal treatment above dealt with as common to all papers.

The Committee have been desirous of bringing their investigations to a practical conclusion in specific terms—viz. by the suggestion of standards of quality. It is evident that in the majority of cases there is little fault to find with the practical adjustments which rule the trade. They are, therefore, satisfied to limit their specific findings to the following—viz. (1) normal standard of quality for book-papers required for publications of permanent value. For such papers they specify as follows:

Fibres: Not less than 70 p.ct of fibres of class A; class D excluded.

Sizing: Not more than 2 p.ct. rosin, and finished with the normal acidity of pure alum; starch excluded.

Loading: Not more than 10 p.ct. total mineral matter (ash).

(2) With regard to written documents, it must be evident that the proper materials are those of class A, and that the [Pg 161] paper should be pure and sized with gelatin, and not with rosin. All imitations of high-class writing-papers which are, in fact, merely disguised printing-papers, should be carefully avoided.

Appendix.—To the Report is added 'Abstracts of Papers' in 'Mittheilungen aus den Koniglichen Technischen Versuchsanstalten, Berlin,' for the years 1885-1896 inclusive—which is, in fact, a summary of the investigations of the Institution in connection with paper and paper-standards.

(p. 273) **Special Industrial Developments.**—From the point of view of the chemist there has been a very large development of the cellulose industries during the last five years. This is not so much marked by the gradual and progressive growth of the well-established industries, as by the success of the newer ones, with the attendant forecast of enormous developments of the industries in artificial products, the manufacture of which rests upon a purely chemical basis. We can, of course, only treat them from this limited standpoint, and so far as they involve and elucidate chemical principles.

I. Chemical Treatments of Raw Materials.

(*a*) **Flax-spinning.**—The treatment of the roving on the spinning-frame by the addition of reagents to the macerating liquid—otherwise and usually hot water—continues to be justified by results. The technical basis of the process and the reactions determined in the spinning-trough by the alkaline salts used—chiefly sulphite and phosphate of soda—is set forth in the original work, p. 280. Since that time a sufficient period has elapsed to judge the effects, both technical and industrial, by the results of a commercial undertaking based on the exclusive use of the process. Such a concern is [Pg 162] the Irish Flax Spinning Company of Belfast. At this mill the experience is uniform and fully established that by means of the process the drawing, *i.e.* spinning, quality of inferior flaxes is very considerably appreciated, enabling the spinner to use such flaxes for yarns of fineness which are unattainable by the ordinary method of spinning through hot water. Notwithstanding the success of this undertaking the development of the method is still inconsiderable. It is none the less a further and forcible demonstration of the existence of margins of increased technical effect which it is the work of the scientific technologist to exploit.

(*b*) **Wood-pulp and Methods of Manufacture.**—There is a steady growth in the consumption of wood-pulps (cellulose) relatively to other materials. In regard to the paper-trade of the world, this continues to be one of the most prominent characteristics of its evolution. In the United Kingdom the conditions of its competition are of a more special kind by reason of the firm foothold of esparto, which is a most important staple in the manufacture of fine printings. Whereas the consumption of esparto remains nearly stationary at about 200,000 tons per annum, the importation of wood-pulps has shown the extraordinary rate of increase of doubling itself every five years. But in the group 'wood-pulps' the trade returns have until recently included the 'mechanical' or ground wood-pulps. From 1898 we have separate returns for the chemical or cellulose pulps, and in 1899 the tonnage reached nearly to that of esparto, with a total money value about 80 p.ct. greater. When it is remembered that this is one of the newer chemical industries in cellulose products, and that these large commercial results have been accomplished during a period of twenty years, we are impressed with the scope of the industrial outlook to the chemist, afforded by the arts of which cellulose is the foundation. [Pg 163]

It may be noted that there have been no important developments in the purely chemical processes involved in the several systems of preparing cellulose from wood. The acid methods (bisulphite processes) have developed much more extensively than the alkaline, the latter including the caustic soda and the mixed sulphide ('Dahl') process. The bisulphite processes depended in the earlier stages upon the efficiency of lead-lined digesters. But the problem of acid-resisting linings has been much more perfectly solved in later years in the various types of cement and other silicate linings now in use. The relative permanency of these linings has had an important effect on the costs of production. Further economies result from the use of digesters of enormous capacity, dealing with as much as 100 tons of wood at one operation. As a combined result of economic production and active competition, the selling prices of 'sulphite pulp' have moved steadily downwards in relation to other half-stuffs and raw materials. As a necessary consequence the prices of those which it has gradually displaced have depreciated, and a study of the price and tonnage-equilibrium as between rags, esparto, and wood-pulp over a series of years forms an interesting object-lesson in the struggle for survival which is an especial mark of modern industry. For these matters the reader is referred to the special literature of the paper-making industry. [12]

It is not a little remarkable that the main by-product of these bisulphite processes—the sulphonated derivatives of the lignone constituents of the wood—is still for the most part an absolute waste, notwithstanding the many investigations of technologists and attempts to convert it to industrial use (see [Pg 164] p. 149). Seeing that it represents a percentage on the wood pulped equal to that of the cellulose obtained, it is a waste of potentially valuable material which can only be termed colossal. Moreover, as a waste to be discharged into water-courses, it becomes a source of burden and expense to the manufacturer, and with the increasing restrictions on the pollution of rivers it is in many localities a difficulty to be reckoned with only by the cessation of the industry. The problem in such cases becomes that of dealing with it destructively, *i.e.* by evaporation and burning. In this treatment the obviously high calorific value of the dissolved organic matter (lignone) appears on the 'credit' side. But where calcium and magnesium bisulphites are used, the residue from calcination is practically without value. It appears, however, that by substituting soda as the base the alkali is recoverable in such a form as to be directly available for the alkaline-sulphide or 'Dahl' process. As a more complicated alternative the soda admits of being recovered on the lines of the old black-ash or Leblanc process, and the sulphur by the now well-established 'Chance' process, for which, of course, an addition of lime is necessary to the fully evaporated liquors previously to calcining. The engineering features of the system, so far as regards evaporating and calcining, are the same. For economic working there is required (*a*) evaporation by multiple effect and (*b*) calcining on the continuous rotary principle. For the latter a special modification has been devised so that the draught of air is concurrent with the movement of the charge in the furnace, securing a progressively increasing temperature within the furnace. This interesting development of the chemical engineering of wood-pulp systems has been elaborated by two well-known technologists, Drewson and Dorenfeldt, and readers who wish to inform themselves in detail of these developments are referred to the various publications of these inventors. [Pg 165]

Assuming the present necessity of a destructive treatment of the by-products of the bisulphite processes, the scheme has many advantages. The soda-bisulphite liquors are more economically prepared; the pulp obtained is superior in paper-making quality to that resulting from the lime or magnesia (bisulphite) processes: it is more economically bleached.

Then, as pointed out, the soda may on the one plan be obtained in a form in which it is immediately available as a powerful hydrolysing alkali in the manufacture of a 'soda' pulp. These two systems become, therefore, in a new sense complementary to one another. Lastly, it is obvious that the employment of soda as the base opens out a new vista for developing the electrolytic processes of decomposing common salt.

The authors have assisted in preparing plans for a comprehensive industrial scheme combining all these more modern developments.

In this scheme it is only the combination which is novel, and as it involves no new principles in the chemical treatments of the materials we are not further concerned with it than to have briefly sketched its economic basis. This may be summed up in result in the important question of cost and selling price, and the estimate is well grounded that by means of this scheme *bleached wood-pulp* can be sold on the English market at 10*l*. a ton. It is important to note this figure and to compare it with the prices of twenty years ago. The fall has been continuous, notwithstanding the influence of the opposing factors of increasing consumption, exhaustion of accessible supply of timber, and relative appreciation of the essential costs of steam, chemicals, and labour. It is important in forecasting the future, since the youngest and apparently most promising of the 'artificial' cellulose industries employs wood-cellulose by preference as its raw material (see p. 173). [Pg 166]

As a last point it must be considered that as chemists we are bound to anticipate the realisation of value in the soluble by-products of the bisulphite processes. Outside the intrinsic interest attaching to the solution of this problem, it carries with it the promise of a further economy in the production of wood-cellulose.

Bleaching of Vegetable Textiles.—By far the largest of these industries are those which are engaged in producing the 'pure white' on cotton and flax goods. The process, considered chemically, is simply that of isolating a pure cellulose, and we endeavoured to give due prominence to this view in the original work. It is important to insist upon it for the reason that this view gives the due proportion of chemical value to the several contributory treatments—alkaline hydrolyses (caustic lime and soda boils), hypochlorite oxidations, and incidental acid treatments (souring). The first of these is by far the largest contributor of 'chemical work,' though the second, by being the agent for the actual whitening effect or bleaching action proper, occupies a position of often exaggerated importance.

In bleaching processes there has been no radical change of system on the large scale since the introduction of the 'Mather' kier in 1885, and the associated change from lime and ash boiling to the caustic soda circulating boil with reduced volume of lye, which this mechanical device rendered practicable. It is outside the scope of this work to follow up this branch of technology in any detail, and we cannot discuss the evolution of systems on variations of detail where no essential principle is involved. But we have to notice a very recent development which has only just begun its industrial career, and which does give effect to a principle of treatment not previously applied. This is tersely stated by its originator, [Pg 167] William Mather, [13] in the expression, 'it is more economical to make liquids pass through cloth than to make cloth pass through liquids.' The starting point of this development is the invention of a complete self-contained machine in which a rolled batch of cloth can receive a succession of chemical treatments, with accessory washings—the solutions, or wash waters, being circulated through the cloth. The essential fact on which this system is based is that a perfect liquid circulation can be maintained from selvedge to selvedge through the folds of a tightly rolled batch of cloth. Such circulation is therefore quite independent of the diameter of the batch. If we consider a cloth under chemical treatment with solutions, it is clear that the reactions and interchanges of soluble matters within the cloth, within the twisted elements of the yarn, and in the last grade of distribution within the actual ultimate fibres, are subject to capillary transmission, and osmotic exchange. There is a mixture of these molecular effects, with the circulation in mass, sweeping both faces of the cloth. It is obvious that for the mass effect a relatively very small volume of circulating liquid is necessary to maintain uniform conditions of action. In the actual disposition of the machine the rolled batch of cloth nearly fills the cylindrical space of what we may call the reaction chamber, and the circulation of the liquid is maintained by a circulating pump and a differential pressure in the horizontal plane across and through the folds of the batch. This is in the meantime kept in slow revolution. For a full description of these mechanical details the reader is referred to the original patent specifications [Engl. Pat. 23,400, 23,401; 1900, W. Mather]. If we again consider the principles involved, they are very much as set forth in our [Pg 168] original work (pp. 288-291). Boiling processes in which a relatively large volume of liquid is used are wasteful of steam, the active agent is unnecessarily diluted or used in superfluous quantity, and the soluble by-products, being continually removed as formed, cannot so effectively contribute by secondary actions to the chemical work. The new mechanical appliance enables us to further reduce the volume of liquid required in the alkaline-hydrolytic treatment of vegetable textiles, and where advantageous to bring the treatment down (or up) to a process of steaming with the active agent dissolved in a minimum proportion of water relative to the cloth. This concentration of effect is of importance in flax cloth, and especially linen treatment, where the peculiarly resistant cutocelluloses have to be attacked and a considerable proportion of waxy by-products to be removed. These points are the basis of the special process of Cross and Parkes [Engl. Pat. 25,076/ 99] for steaming flax (and cotton) goods with an emulsion containing, in addition to the special hydrolysing agent—caustic soda—mixtures of soap with 'mineral' or other oils, the presence of which effectually aids the removal of the by-products in question.

A complete system on these lines is now working on the industrial scale in the Belfast district. The results are not merely economical in largely reducing the number of alkaline boiling treatments required on the old plan of pan or 'pot' boiling, but are visible in the strength and finish of the linens so treated.

For cotton bleaching the costs may be put down at a fraction of those of the Irish linen bleach. The economical advantages of the new system are obviously less in relation to the lesser total costs. But there are other points which have come into more prominent influence. The mechanical wear and tear on the cloth is considerable in the ordinary process, more [Pg 169] especially in the mangle-washes. As a result the adjustment of warp and weft is more or less disturbed. These defects are absent from a system which operates on the cloth in a fixed position.

But as we are mainly concerned with the purely chemical factors we cannot pretend to deal with textile questions. We have to notice the remaining element of chemical economy as it involves a fundamental principle. The practice of washing residues or products of reaction free from reagents and soluble by-products involves a well-known mathematical law, under which the rate of purification is a function rather of the *number* of successive changes of washing liquid than of the volume of the latter. The ordinary practice of textile washings entirely ignores this principle, and the consumption of water in consequence may reach many thousand times the economic minimum. With supplies of water often in indefinite excess of requirements, even in this most wasteful method, bleachers are in no need to consider the question of consumption. But leaving aside particular and local considerations of advantage the fact is that the new system gives control of the practice of washing, enabling the operator to adapt an important element of the daily routine to a fundamental principle which has been almost universally ignored.

In the oxidising processes which follow the alkaline treatments, the hypochlorites are still the staple agents. Owing to the steady relative fall in the selling prices of the permanganates these are coming into more extensive use, but the consumption is still small, and they are mainly used for certain special effects, chiefly in linen or more generally flax cloth bleaching.

Paper-pulp Spinning.—Paper is a continuous web or fabric produced by the interlocking of the structural fibrous units of the well-known short length. In Japan and other [Pg 170] countries paper is made to serve for all or some of the purposes for which we employ string or twine, and to give the necessary tensile strength the paper is twisted or rolled on itself. Such twisting, however, adds nothing to the intrinsic tensile qualities of the original paper.

A new technical effect is realised in this direction by the treatment of paper-pulp in the process of its conversion into a continuous web: The pulp is formed into continuous strips of convenient breadth (usually from 2 to 8 mm.), these receive a 'rolling-up' treatment immediately following the squeeze of the press rolls by which the superfluous water is removed: they are then further but incompletely dried, and in this condition are subjected to a final spinning or twisting treatment on ring-spinning machinery of special construction.

Such a process was originally patented by C. Kellner in this country (E.P. No. 20,225/1891), and is fully described in his specification. Later improvements in detail were patented by G. Türk (E.P. 4621/1892).

A joint system is now being industrially developed in Germany by the Altdamm-Stahlhammer Pulp and Paper Company under the technical direction of Dr. Max Müller, and there appears to be every prospect of the product taking a position as a staple textile.

The process has only the incidental interest in connection with our main subject, that it employs chiefly the 'chemical' pulps or celluloses as raw materials. The industrial future of the application must, of course, be largely determined by costs of production, as the directions of application in the weaving industries will be limited by the necessarily inferior grade of tensile strength belonging to these products and the degree by which this is lowered on complete wetting. All these questions have been duly weighed by those engaged in this interesting development, and the conclusion of those [Pg 171] qualified to judge is that the new industry has vindicated for itself a permanent position.

II. The Chemical Derivatives of Cellulose, in their industrial aspects, have come to occupy a profoundly important position in the world's affairs. In the way of any essential alteration of the perspective from that obtaining in 1895 we have nothing to chronicle. No new derivatives of industrial importance have been added in that period; but certain new methods incidental to the preparation of well-known compounds or for converting them into more generally available forms have been introduced, and these are contributing to the rapid expansion of the 'artificial' cellulose industries.

Of the cellulose esters the nitrates are still the only group in industrial use. There uses for explosives have attained immense proportions, and their applications for structural purposes are continually on the increase. The manufacture of smokeless powders on the one hand, and of celluloid and xylonite (both in the form of films and solid aggregates) on the other, has taken no new departure. The industry in 'artificial silks' or 'lustra-celluloses,' by the collodion processes also, whilst presenting features of unusual interest attaching to rapid expansion, has been barren of contribution of fundamental scientific or technical importance. The tetracetate is now manufactured on the large scale, but the product has yet to make its market.

The process of mercerising cotton yarns and cloth has been developed to an industry of colossal dimensions, and the growth has been especially rapid during the last five years. Significant of the technical progress in these two industries, with their common aim of appreciating cellulose in the scale of textiles by approximating its external properties in those of silk, is the appearance of a monograph of the technology of each, notices of which have been previously given (pp. 22-26).

There is little doubt, however, that the question of the [Pg 172] future industry in the various forms of cellulose, thread, film, structureless powder or solid aggregate, obtainable by artificial means, mainly turns upon cost of production. Irrespective of cost, there would, no doubt, be a market for all these products, based upon such of their properties or effects as are indispensable and not otherwise obtainable. As an illustration, we may cite the extraordinary selling prices of 40-50 fr. per kilo, for the 'artificial silks' (collodion process) which ruled some three years ago; and we may note that for a special application of viscose the dissolved cellulose is paid for at the rate of 10s. per lb. These facts are certainly worthy of mention, and should be borne in mind as an index of some special features of modern manufacturing industry. But with a material like cellulose rendered available in a new shape the question which always arises more prominently than that of limited uses at high prices is that of consumption on the extensive scale which marks the older and well-known products. That question is rapidly solving itself in this country as regards the 'artificial silks.' There is at present a limited market at 9s.-10s. per lb., a price which on the one side excludes extensive consumption, and on the other practically bars manufacture in this country by any of the collodion systems. It will appear from a very elementary calculation of what we may call the theoretical costs that the above selling price would not have a remunerative margin. The theoretical costs are made up of

[Pg 173]

Raw materials [14]	Cotton. Nitrating acid. Ether-alcohol (solvent). Denitrating chemicals.
Labour	(a) Nitrating and preparing collodion. Denitrating and bleaching. (b) Textile operations. Spinning. Winding and twisting. Rewinding.

| Power | Making, filtering, and distributing collodion. Driving textile machinery. |

Added to which are the costs of expert management and supervision and general establishment expenses. It is evident that raw materials make up a large fraction of the total cost; also that a very large item is the waste work of converting the cellulose into nitrate, only to remove the nitric groups so soon as the cellulose is obtained as thread.

It is clear that the aqueous solutions of cellulose have a double advantage in this respect — not only do they readily yield an approximately pure cellulose as a direct product of regeneration or decomposition, but the first cost of the solution is very much less. With these newer products, therefore, the spinning problem enters on a new phase of struggle. It is certain that at selling prices at or about 5*s.* to 7*s.*, very large markets will be open to the product or products. The two processes which are or may be able to fulfil this demand are those based (1) on cuprammonium solutions of cellulose, (2) on the sulphocarbonate or viscose. As regards *first cost* of the solution the latter has a large advantage. One ton of wood pulp (at 12*l.*) can certainly be obtained in solution in a condition ready for spinning at a total cost (materials) of less than 30*l.* The cuprammonium process, so far as 'outside' information goes, requires for production of the solution (1) cotton as raw material, (2) ammonia (calc. as concentrated aqueous) equal to 1-1/2 times its weight, and (3) metallic copper 25 p.ct. of its weight; and the costs are approximately 100*l.* per ton. It is obvious that the materials are recoverable from the precipitating-bath, but at a certain added cost. We have no statements as to the proportion recoverable nor the costs incurred, and we are therefore unable to measure the total net cost of the regenerated cellulose by this process. It is certainly much less than by the collodion processes. As to the textile quality of the thread, the product has not yet been on a sufficiently wide selling basis for that to have been determined. There are a great many factors which enter here. Not merely the external characters of lustre, softness, and translucency, but the all-important quality of uniformity of thread. The collodion-spinning is a process still very defective in this respect, and the defect is no doubt referable to the difficulty of securing absolute physical invariability of the collodion. It is to be regretted, in the interests of scientific development, that none of the technologists who have published investigations of these processes have entered into the discussion of the fundamental factors of the spinning processes; we are, therefore, unable at this stage to discuss these elements of a full comparison in greater detail. We cannot, for this reason, say how far the cuprammonium process diverges in point of control from the standard of the collodion processes. Of the 'viscose' product we have a more intimate knowledge, and it certainly reaches a higher general standard than the older and now well-known artificial silks. The process is also sufficiently developed to enable the total costs of production to be estimated at a figure less than one-half that of the 'collodion' processes. This would assure to this system an *entrée* in this country, and a basis of expansion limited only by the ordinary laws of supply and demand.

This prospect is opened up precisely at the moment when, for various reasons connected both with the difficulties of [Pg 175] manufacture and the narrowing of the margin of profit, the proprietors of the two systems of collodion-spinning have decided to abandon all idea of manufacturing by these systems in this country. [15] We leave the discussion of the industrial problem at this point.

In regard to other developments based upon the exceptional character and properties of the sulphocarbonate, their further discussion will exemplify no general principles; and as regards technical detail they have been dealt with in the papers previously noticed.

As a purely general question, if there is to be any industry in these 'artificial' forms of cellulose, commensurate with the magnitude that usually belongs to the cellulose industries, it must come by way of a plastic or soluble form prepared at low cost, and conserving the essential molecular properties of the cellulose aggregate. These are the particular features of the sulphocarbonate. The obvious difficulties in the way of its industrial applications are those caused by the presence of alkali and sulphur compounds. These are dealt with by appropriate chemical means; but the fact that there is a special chemistry of the product has rendered its industrial progress slow. The work of the last five years in this, as in other applications of cellulose in its many derived forms, has resulted in a considerable addition to the domain of practical chemistry.

Further developments will make an increasing demand upon our grasp of the fundamental constitutional problems, to which it is the main purpose of the present volume to contribute.

FOOTNOTES:

[11] This is the most complete notice that has appeared and the bibliography is exhaustive. The publication comes into our hands too late to be noticed in detail.

[12] *Text-book on Paper-making*, Cross and Bevan (Spon, London: second edition, 1900). *Chemistry of Paper-making*, Griffin and Little (New York, 1894: Howard Lockwood & Co.). *Handbuch d. Papierfabrikation*, C. Hofmann (Berlin). *Paper Trade Review*, London (weekly). *Papier-Zeitung*, Berlin.

[13] William Mather, M.P., of the firm of Mather & Platt, Limited, Manchester.

[14] The actual costs varying considerably in the various countries, we cannot make any specific statement. But from estimates we have made, the costs of obtaining cotton in filtered solution as collodion multiply its value by 12-14, the denitrations adding further costs and raising this multiple to 18-20. In the same estimates we arrived at the conclusion that the item for raw materials made up 60 p.ct. of the total cost of the yarn.

[15] The recent failure of a French company founded for the exploitation of the cuprammonium process may be taken as showing that it presents very considerable technical difficulties. It is a matter of common knowledge that this company *estimated* the costs of production to be such as to enable the product to be sold at 12 fr. per kilo., whereas the costs actually obtaining were a large multiple of this figure.